APPLIED MINERALOGY
IN THE MINING INDUSTRY

APPLIED MINERALOGY
IN THE MINING INDUSTRY

William Petruk
Ottawa, Ontario, Canada

2000
ELSEVIER
Amsterdam - Lausanne - New York - Oxford - Shannon - Singapore - Tokyo

ELSEVIER SCIENCE B.V.
Sara Burgerhartstraat 25
P.O. Box 211, 1000 AE Amsterdam, The Netherlands

First edition 2000

Library of Congress Cataloging-in-Publication Data

Petruk, W., 1930-
    Applied mineralogy in the mining industry / William Petruk.-- 1st ed.
      p. cm.
    Includes bibliographical references and indexes.
    ISBN 0-444-50077-4 (alk. paper)
      1. Mineralogy, Determinative. 2. Mines and mineral resources. I. Title.

    TN260 .P485 2000
    549.1--dc21

                                                    00-062224

ISBN: 0 444 50077 4

♾ The paper used in this publication meets the requirements of ANSI/NISO Z39.48-1992 (Permanence of Paper).
Printed in The Netherlands.

PREFACE

An evolution in applied mineralogy related to processing began in the mid 1960's, and significantly modified the scope of the field. Many new instruments were developed and enabled applied mineralogists to *accurately* determine a wider array of mineral characteristics than was previously possible. The results had a major impact on processing, and many operations now use applied mineralogy as a standard tool. The expanded scope of applied mineralogy made one realize that a dedicated mineralogist is essential at most large mining operations; a person who performs applied mineralogy as an adjunct to other duties can no longer develop the breadth of knowledge and experience that is required to recognize and determine mineral properties that affect different processes. It is also recognized that, to be able to assess what to look for, the applied mineralogist should be a member of a team performing investigations.

When I joined the Mineralogy Section, Mineral Sciences Division, Mines Branch, Department* of Mines and Technical Surveys as a mineralogist in 1960, the stature of the field of applied mineralogy was only beginning to develop. I was in the right place to participate. The organizational structure of the department, the highly qualified and motivated colleagues, and the timing were perfect for my involvement in the evolution of this fascinating field.

During the period of about 1960 to 1970, there was a dynamic desire among the scientists in the Mineral Sciences Division to improve techniques of performing mineralogical analyses. Thus, when the microprobe was developed in France, our Division undertook to build a microprobe before reliable commercial ones were available. The project was eventually scrapped.

The emergence of the microprobe was the first in a series of appearances of new instruments that could be used to characterize minerals. The new instruments enabled applied mineralogists to determine standard mineralogical data quickly and more accurately than was previously possible, to acquire data that previously could not be obtained and to solve processing problems where, previously, the solutions could only be extrapolated. However, to determine appropriate mineral characteristics, research was required to adapt and modify the available instruments. To cope with this new era of applied mineralogy, each scientist in the Mineralogy Section focussed on a specific area of research.

My main assignment was to support the work in other Sections and Divisions in the Department by characterizing ores, materials and process products. A secondary responsibility was to perform research when the support work was completed. Most, but not all, of my assignments and research were related to mineral processing. During the period of 1960 to 1972 the Mines Branch employed many mineral processing engineers who performed laboratory and

---

In 1960 the area of work was named the Mines Branch, Department of Mines and Technical Surveys. The work area was renamed to CANMET, Department of Energy, Mines and Resources in 1974, and to CANMET, Department of Natural Resources Canada in 1995.

pilot plant tests on a wide variety of ores. I characterized many of these ores and process products. The mineral processing engineers frequently asked for more information than could be determined by the techniques that existed at the time. Dan Pickett, one of the senior Mineral Processing engineers in the Mines Branch, often commented that *"complete mineral characterization solves most mineral processing problems"*.

Subsequently, the new instruments made it possible to obtain additional data. However, to provide full support, and to adapt the new instruments to determine mineral characteristics that helped the mineral processing engineers solve mineral processing problems, a basic understanding of mineral processing fundamentals and techniques was necessary.

An instrument that showed promise for solving mineral processing problems was an image analyser. I started using the instrument when the first reliable model was developed in 1972. During the period of 1979 to 1981, I had the good fortune of holding many discussions, with Dr. A. Read of CSIRO, who headed the group that developed the QEM*SEM for determining mineral quantities and mineral liberations. Our department did not have the funds for such sophisticated equipment. So, with support of Dr. P. Mainwaring, I developed the MP-SEM-IPS by integrating a microprobe, energy dispersive X-ray analysers (EDS) and image analyser using special software. The MP-SEM-IPS performs essentially the same analyses as the QEM*SEM, but in a different manner, and requires a specialized operator. On the other hand the MP-SEM-IPS has additional image analysis capabilities, and Dr. R. Lastra is continuing the research at CANMET to improve its applications.

The accurate data that was obtained with image analysers emphasized the problem of measuring mineral liberations observed in polished sections, and brought about a new challenge to researchers world-wide, including Prof. P. King, Salt Lake City, Utah, USA, the late Prof. G. Barbery, Laval University, Québec City, Canada, Dr. R. Lastra, CANMET, myself and many others too numerous to cite. The challenge is on-going.

As the benefits of mineral characterization became apparent, and as mineralogy became a significant if not a necessary step in most processes, professional organizations were formed worldwide. By 1980, the applied mineralogists in North America, including Drs. D.M. Hausen, R.D. Hagni and W.C. Park, formed an Applied Mineralogy Committee within two Divisions of the American Institute of Mining, Metallurgical and Petroleum Engineers, Inc. In the meantime the applied mineralogists in South Africa, including Drs. S.A. Heimstra, S.A. De Waal, Mr. L.F. Haughton and others, formed an international group, ICAM (International Congress on Applied Mineralogy). I was fortunate to actively participate in both organizations, and to co-edit proceedings from six symposia and congresses.

The purpose of this book is to share what I learned about the role, applications and benefits of applied mineralogy. The book covers three aspects of applied mineralogy. The first part (Chapters 1 and 2) discusses methods of performing applied mineralogy studies, and capabilities and applications of newly developed instruments. The second part (Chapters 3 to 9) describes studies, findings and interpretations related to commodities, as examples of the types of information that can be obtained, and of problems that can be solved, by applied mineralogy. The chapters on commodities cover studies on base metal (copper, zinc and lead) volcanogenic ores, gold ores, porphyry copper ores, iron ores, and industrial minerals. The third part (Chapter 10) discusses the characteristics of tailings from mineral processing operations, with emphasis on related environmental problems and on remediation.

William Petruk, PhD

ACKNOWLEDGEMENTS

The author gratefully acknowledges the persons who reviewed various chapters of the book and made significant improvements to the contents. The reviewers were:

R.E. Healy, Minoretek, Winnipeg, Manitoba, Canada,
Dr. K. Kojonen, Geological Survey of Finland, Espoo, Finland,
Dr. R. Lastra, CANMET, Department of Natural Resources Canada,
Dr. D. Leroux, Centre de Technologie Noranda, Pointe Claire, Québec, Canada,
Jan Nesset, Centre de Technologie Noranda, Pointe Claire, Québec, Canada,
Staff, Centre de Technologie Noranda, Pointe Claire, Québec, Canada,
Dr. D.B. Sikka, Cabinet conseil en Geologie Minière Sikka Enr., 2108, Montreal, Québec, Canada,
Dr. J. A. Soles, Ottawa, Ontario, Canada,
Dr. J. Wilson, CANMET, Department of Natural Resources Canada,
E. vanHuyssteen, CANMET, Department of Natural Resources Canada,

Grateful acknowledgements are extended to:
- R.E. Healy, Minoretek, Winnipeg, Manitoba, who studied the Flin Flon and Snow Lake ores under a Canada Mineral Development agreement. A summary of the results and photomicrographs taken by R.E. Healy constitute Chapter 4 of this book.

- H. Mani, INCO, Mississauga, Ontario, Canada, who was the leader of a project under a Canada Mineral Development agreement. Ms. Mani supervised an investigation on pentlandite flotation from a serpentinized ore, and performed a mineralogical study on the ore. The results are summarized in the last part of Chapter 5, and a photomicrograph taken by Ms. Mani is included in the chapter.

- L. Lewczuk, who studied samples at New Brunswick Research and Productivity Council, Fredericton, New Brunswick, Canada, under a Canada Mineral Development agreement. The samples were from the grinding and spiral circuits of the Scully concentrator, Wabush Mines Limited. Some of the information and a few of the photomicrographs taken by Mr. Lewczuk were used in Chapter 8.

Thanks are extended to Larry Urbanoski, Brunswick Division, Noranda Mines Limited, Bathurst, New Brunswick, and to Chris Larsen, Centre de Technologie Noranda, Pointe Claire, Québec, Canada, for permission to use the information that is included in the last part of Chapter 3. The information was taken from reports of a 1993 to 1995 task force study on the feed, concentrates and tailings from flotation cells in the Copper rougher circuit of the Brunswick concentrator.

CREDITS

The following figures were reproduced from figures in published papers.

Figure 1.5 (a, b, c), reproduced from Figures 4.2b, 4.2c 4,2e, in PhD thesis by D. Lin, Department of Mining and Metallurgical Engineering, McGill University. Reproduced with permission of the Department.

Figure 1.6 (a, b), reproduced from Figures 2 and 3 in the paper by R.G. Fandrich, C.L. Schneider and S.L. Gay, Two Stereological Correction Methods: Allocation Method and Kernel Transformation Method, Mineral Engineering, 11, 8 (1998). Reproduced with permission of Elsevier Science B.V.

Figure 2.4, reproduced from Figure 2 in paper by J.L. Jambor, A.P. Sabina, R.A. Ramik, and B.D. Sturman, A fluorine-bearing gibbsite-like mineral from the Francon Quarry, Montreal, Quebec, Can. Mineral., 28,2 (1990). Reproduced with permission of J.L. Jambor.

Figure 6.3 and 6.4, reproduced from Figures 2b and 2c in paper by J. Guha, A. Gauthier, M. Vallee and F. Lange-Brard, Gold mineralization patterns at the Doyon Mine (Silverstack), Bousquet, Quebec, Geology of Canadian Gold Deposits, CIM, Spec. Vol. 24 (1982). Reprinted with permission of the Canadian Institute of Mining, Metallurgy and Petroleum.

Figures 6.7, 6.8, 6.9, 6.13, 6.14, 6.15, and 6.16, Reproduced from figures in paper by K. Kojonen, B. Johanson, H.E. O'Brien and L. Pakkanen, Mineralogy of Gold Occurrences in the Late Archean Hattu Schist Belt, Ilomantsi, Eastern Finland, *Geol. Survey of Finland*, Special paper 17 (1993). Reproduced with permission of K. Kojonen.

Figure 6.20, reproduced from Figure 10 in paper by D.M. Hausen, Process Mineralogy of Auriferous Pyritic Ores at Carlin, Nevada, in Process Mineralogy (1981). Reproduced with permission of TMS.

Figures 7.1 and 7.2, reproduced from figures 5 and 9 in paper by M.J. Osatenko and M.B. Jones, Valley Copper, in Porphyry Deposits of the Canadian Cordillera, CIM Spec. Vol. 15 (1976). Reprinted with permission of the Canadian Institute of Mining, Metallurgy and Petroleum.

Figure 10.6, reproduced from Figure 10 in manual by E. vanHuyssteen, Overview of Acid Mine Drainage in the Context of Mine Site Rehabilitation, CANMET/MMSL, Dept. of Natural Resources Canada (1998). Reproduced with permission of E. vanHuyssteen.

TABLE OF CONTENTS

CHAPTER 1

GENERAL PRINCIPLES OF APPLIED MINERALOGY

CHAPTER 2

INSTRUMENTS FOR PERFORMING APPLIED MINERALOGY STUDIES

CHAPTER 3

MINERALOGICAL CHARACTERISTICS AND PROCESSING OF MASSIVE SULFIDE BASE METAL ORES FROM THE BATHURST-NEWCASTLE MINING AREA

CHAPTER 4

VOLCANOGENIC BASE METAL DEPOSITS IN THE FLIN FLON-SNOW LAKE AREAS, MANITOBA, CANADA

CHAPTER 5

RELATIONSHIPS BETWEEN MINERAL CHARACTERISTICS AND FLOTABILITY

CHAPTER 6

APPLIED MINERALOGY RELATED TO GOLD

CHAPTER 7

APPLIED MINERALOGY: PORPHYRY COPPER DEPOSITS

CHAPTER 8

MINERALOGICAL CHARACTERISTICS AND PROCESSING OF IRON ORES

CHAPTER 9

APPLIED MINERALOGY INVESTIGATIONS OF INDUSTRIAL MINERALS

CHAPTER 10

APPLIED MINERALOGY TO TAILINGS AND WASTE ROCK PILES - SULFIDE OXIDATION REACTIONS AND REMEDIATION OF ACIDIC WATER DRAINAGE

CHAPTER 1

GENERAL PRINCIPLES OF APPLIED MINERALOGY

1.1. INTRODUCTION

Applied mineralogy in the mining industry is the application of mineralogical information to understanding and solving problems encountered during exploration and mining, and during processing of ores, concentrates, smelter products and related materials. It involves characterizing minerals and materials, and interpreting the data with respect to (1) exploration, (2) mineral processing, (3) tailings disposal and treatment, (4) hydrometallurgy, (5) pyrometallurgy, and (6) refining. Many techniques for determining mineral characteristics have been developed in the last three decades as a result of the availability of new equipment which includes scanning electron microscope equipped with an energy dispersive X-ray analyser (SEM/EDX), environmental scanning electron microscope (E-SEM), microprobe (MP), image analyser (IA), proton-induced X-ray analyser (PIXE), secondary ion mass spectrometer (SIMS), time of flight-secondary ion mass spectrometer (ToF-SIMS), laser ionization mass spectrometer (LIMS), time of flight-laser ionization mass spectrometer (ToF-LIMS), infra-red analysis (IRA), cathodluminescence and others. The results obtained by using the above-mentioned equipment have increased our knowledge of mineral characteristics and have provided a better understanding of mineral behaviours during processing.

Mineral exploration techniques include identifying minerals and determining mineralogical characteristics of rocks and ore deposits. The results of mineralogical studies related to exploration are often used to predict (1) locations of ore deposits, (2) the potential for recovering certain minerals, metals or elements, and (3) the behaviour of ores during mineral processing.

Mineral processing operations are performed to produce concentrates that have satisfactory grades and metal or mineral recoveries, and tailings that can be disposed of in an environmentally safe manner. The operations are generally evaluated by chemical assays of the products, and by materials balancing of the assay results. The evaluations made from assay results need to be augmented by mineralogical data when processing problems are due to mineralogical characteristics of the ore and/or process products. In some instances an experienced microscopist can identify the processing problem at a glance, but, generally, extensive mineralogical investigations need to be performed.

Tailings are studied to determine whether more minerals can be economically recovered, and to evaluate mineral reactivities which may produce and release elements that are deleterious to the environment. The studies are performed by identifying the minerals and determining mineral compositions and textures in fresh and altered tailings.

Hydrometallurgy operations (including cyanidation of gold) involve leaching materials to produce a leach liquor and a residue. The products are evaluated by chemical analysis. The behaviour of the materials during leaching and the characteristics of the leach residues are determined by applied mineralogy techniques (Chen and Dutrizac, 1990).

Pyrometallurgical operations involve smelting of concentrates and other products to produce metals and slags. The purity of the metals, the losses to the slag, and the environmental conditions in the workplace are generally evaluated by chemical analysis and materials balancing. The characteristics and quality of the metals are usually determined by microstructural characterization techniques. The characteristics of other smelter products and slags, the reasons for losses to the slags, and the characteristics of airborne dusts in the work place are defined by applied mineralogy techniques which involve using equipment such as optical microscopes, SEM, MP, XRD, cathodoluminescence and image analysis (Craig et al., 1990; Fregeau-Wu et al., 1990; Isaacson and Seidel, 1990; Jokilaakso et al., 1990; Karakus et al., 1990a; Karakus et al., 1990b; Lastra et al., 1998; Mavrogenes and Hagni, 1990; Mavrogenes et al., 1990; Petruk et al., 1984; Petruk and Skinner, 1997; Zamalloa et al., 1995).

Electrolytic refining is performed to increase the purity of products generated by smelting. The smelted product is the cathode in the electrolytic cells and the refined product is the anode. During refining, impurities from the cathode fall to the bottom of the electrolytic cells as residues. The characteristics of the residues and impurities in the anodes are determined by applied mineralogy techniques (Chen and Dutrizac, 1990).

1.2. APPLIED MINERALOGY INVESTIGATIONS

The mineral characteristics that need to be determined by applied mineralogy investigations are:
(1) identities of major, minor and trace minerals,
(2) compositions of minerals that bear on the process,
(3) quantities of minerals,
(4) particle and grain size distributions and textures of the minerals,
(5) mineral liberations,
(6) surface coatings on minerals,

The applied mineralogy investigations are performed by studying:
- uncrushed samples of rocks, ores, drill cores, pyrometallurgical products, anodes, etc.,
- crushed and ground materials from laboratory tests, pilot plant tests, concentrator, tailings piles, hydrometallurgy residues, refinery residues, pyrometallurgical powders, and airborne dusts.

Specialized sample preparation routines are required to prepare the materials for analysis by the different techniques (Stanley and Laflamme, 1998; Nentwich and Yole, 1991). The uncrushed materials are prepared as polished sections, thin sections and/or polished-thin sections, and the crushed and ground materials are prepared as polished sections and/or polished-thin sections. The crushed and ground materials are commonly sieved, and sometimes pre-concentrated by gravitational and other techniques, prior to preparation of polished and/or polished-thin sections. The sections are prepared from the crushed and ground materials by mixing the powders or chips with a resin, allowing the mixture to cure, and preparing sections from the hardened mixture. The polished, thin and polished-thin sections are analysed by optical microscopy, SEM, environmental SEM, MP, IA, SIMS and cathodoluminescence. Analyses by XRD, environmental SEM and IRA are performed on powders, and on individual particles in appropriate mounts. Analyses by LIMS, and TOF-LIMS are performed on individual particles which have been hand-picked and mounted on indium. Airborne dusts are commonly collected on polycarbonate membrane filters or glass slides and are analysed by optical microscopy, XRD, IA, SEM, and MP (Knight et al., 1974; Petruk and Skinner, 1997). When rare minerals are studied, a search for grains of rare minerals needs to be performed.

1.2.1. Identifying minerals and determining mineral compositions

Ideally, all the minerals in a sample should be identified because it is not known which will affect the process. Furthermore, the compositions of some minerals and/or phases need to be determined to define the average quantities of minor and trace elements that may have a bearing on the process.

When applied mineralogy is performed with respect to processing (e.g. mineral processing, hydrometallurgy, pyrometallurgy, refining, etc.) the minerals of economic value need to be identified to guide the engineer in selecting a concentrating process, and the gangue minerals need to be identified because some may interfere with the selected process. In addition, it is necessary to identify minerals that contain trace elements that have a bearing on recovery of the elements and on the purity of refined metals. It is also necessary to identify minerals that release hazardous materials and elements into the workplace and/or the environment during processing.

The main methods of mineral identification include optical microscopy, SEM/EDX, MP and XRD, but other techniques such as cathodoluminescence and IRA are used in some instances. Minor element contents (>200 ppm) in minerals are determined with a MP and trace element contents (<200 ppm) with PIXE, SIMS and/or LIMS.

It is time consuming to identify all the trace minerals as well as minerals that do not have a bearing on the process. Hence information on ore types and process products may suggest mineral assemblages and pinpoint mineral varieties that should be looked for. The minerals that should be looked for when performing applied mineralogy with respect to mineral processing are discussed here for base metal ores, granite hosted tin-tungsten ores, porphyry copper ores and iron ores, as examples.

1.2.1.1. Base metal ores

Base metal ores are processed by flotation to produce Cu, Pb, and Zn concentrates, and the concentrates might recover some Ag, and possibly Au, Cd, Sn, and In. Deleterious elements are As, Sb, Bi, Hg, Se and Te. The mineralogist must, therefore, look for minerals that contain these elements and pay particular attention to mineralogical characteristics that may have a bearing on the process. In addition, some minerals such as galena, chalcopyrite, sphalerite, pyrite, arsenopyrite and tetrahedrite-friebergite-tennantite need to be analysed to determine the minor and trace element contents. The minerals or groups of minerals that need to be identified, and if necessary analysed, when characterizing base metal ores are:

4

(1) *Main minerals of economic value and gangue*: The main minerals of economic value are sphalerite, galena, chalcopyrite, tetrahedrite-friebergite-tennantite and rarely bornite and cassiterite. The main gangue minerals are pyrite, quartz, chlorite, and sericite. Minor and trace gangue minerals are carbonate minerals, other silicate minerals (e.g. amphiboles, pyroxenes, feldspars), barite, arsenopyrite, monazite, titanite and zircon.

(2) *Silver minerals*: The main silver-bearing mineral in base metal ores is tetrahedrite-friebergite-tennantite (Petruk and Wilson, 1993). A significant portion of the silver also occurs as a trace element in galena, and some silver occurs in a variety of silver-antimony-sulfide minerals, including pyrargyrite (Ag_3SbS_3), stephanite (Ag_5SbS_4), miargyrite ($AgSbS_2$), acanthite (Ag_2S), andorite ($PbAgSb_3S_6$), owyheeite ($Ag_2Pb_7(Sb,Bi)_8S_{20}$), jalpaite (Ag_3CuS_2), pyrostilpnite (Ag_3SbS_3) and diaphorite ($Pb_2Ag_3Sb_3S_8$) and possibly other Ag-bearing minerals (Chen and Petruk, 1980; Chryssoulis et al., 1995). In addition, small amounts of silver occur as a trace element in sphalerite, pyrite and chalcopyrite. For a complete evaluation of the mode of occurrence of silver it is necessary to determine the distribution of silver among the different minerals (e.g. mineral balance for silver). This involves identifying the silver-bearing minerals, determining the average amounts of silver in the tetrahedrite-friebergite-tennantite, galena, sphalerite, pyrite and chalcopyrite, and determining the quantities of each mineral. The average amount of silver in the tetrahedrite-friebergite-tennantite is determined with a microprobe, and in the other sulfides with a PIXE, SIMS and/or LIMS. The average quantities of minerals are determined with an image analyser or by point counting.

(3) *Sphalerite*: Sphalerite needs to be analysed to determine the average quantities of Fe, Cd, Mn, Sn, In, Ge and Ag in solid solution. Fe, Cd and Mn are common constituents in sphalerite and their quantities can be determined with an electron microprobe. The quantities of other elements can be determined with a PIXE. Ag in sphalerite from volcanogenic base metal ores varies from <1 to about 100 ppm (ave. ~ 25 ppm), whereas sphalerite from Mississippi-type ores contains up to 500 ppm Ag. The Zn concentrate from the Nanisivik deposit, which is a high grade zinc-lead deposit in northern Canada, contained an average of 270 ppm Ag and 200 ppm Ge, and accounted for ~87 % of the silver in the ore (Bigg, 1980). Cabri et al (1985) analysed 3 sphalerite grains from the Nanisivik ore by PIXE and obtained an average of 670 ppm Ag. One of the grains contained 149 ppm Se and 86 ppm Ge.

(4) *Gold*: Gold in base-metal ores tends to occur as discrete gold grains associated with pyrite, chalcopyrite, arsenopyrite and chlorite, and as invisible gold in pyrite and arsenopyrite. The discrete gold grains float readily in the Cu flotation cells and are recovered in the Cu concentrate. If the gold is present in very small quantities, a gold search which involves scanning the sample needs to be employed to find and analyse the gold grains. The sample can be scanned manually or automatically with an image analyser interfaced to a SEM and EDX.

The quantity of invisible gold in pyrite and arsenopyrite can be determined with a SIMS and/or modern microprobe. Chryssoulis and Cabri (1990) found that pyrite can contain more than 132 ppm invisible Au and arsenopyrite more than 15,200 ppm invisible Au. The quantities of Au in pyrite and arsenopyrite from base metal ores are variable, but are generally much lower. For example, the pyrite in the ore from the Trout Lake base metal deposit in Flin

Flon, Manitoba, Canada, contains from a few ppb to 6 ppm invisible Au (ave ~0.7 ppm) and the arsenopyrite contains from 2 to 130 ppm (ave. ~32 ppm) invisible Au (Healy and Petruk, 1990d). The pyrite in the Mobrun base metal deposit in Quebec, Canada, contains up to 12 ppm Au with highest frequency at 1.5 ppm (Larocque et al., 1995; Petruk et al.,1995).

(5) *Tin and indium minerals*: The tin in volcanogenic base metal ores tends to occur as minute cassiterite inclusions in sphalerite, and to small extent as stannite. Trace amounts of stannoidite, mawsonite, pabstite have also been found in base metal deposits, but they do not account for significant proportions of the tin. Some tin may also occur in solid solution in chalcopyrite, sphalerite and pyrite. Indium tends to occur as a trace element in sphalerite and chalcopyrite (Cabri et al., 1985).

(6) *Secondary copper minerals*: Secondary copper minerals such as covellite, chalcocite, malachite, etc. need to be identified because they interfere with separation of sulfide minerals (McLean, 1984). Quantities as low as 0.1 wt % secondary Cu minerals release enough Cu ions into solution to prevent production of separate Cu, Pb and Zn concentrates of acceptable grade (McTavish, 1985). NaCN may by used to depress small amounts of Cu ions, but not when the ore has large amounts of secondary Cu minerals because too much NaCN will be required to neutralize the Cu ions (Stemerowitz, 1983). Secondary copper minerals such as covellite and chalcocite can be easily recognized with an optical microscope, whereas secondary minerals such as malachite have to be identified by other methods such XRD or SEM/EDX.

(7) *Anglesite (PbSO₄)*: Anglesite needs to be identified because it retards flotation of particles coated with the mineral (Sui et al., 1996). Its presence indicates that the galena has been oxidized and the surface characteristics of the galena have been altered. Its presence also suggests that oxidation of galena may have occurred during mineral processing and released Pb ions which coat other minerals (e.g. sphalerite and pyrite) and affect flotation of these minerals (Chryssoulis et al., 1995; Kim et al., 1995). It is difficult to identify anglesite by standard mineralogical techniques because the mineral occurs as a thin film on the surfaces of other minerals. The thin film can, however, be detected by infrared analysis (Sui et al., 1996) and LIMS, and individual anglesite grains can be identified by XRD or SEM/EDS.

(8) *Deleterious elements*: The quantities of deleterious elements in base metal ores are generally low, but even small quantities may create a problem. For example, small amounts of Se and/or Te in refined copper reduce its quality. The main deleterious elements in base metal ores are:

- Arsenic: Arsenic occurs mainly as a constituent of arsenopyrite and as a trace element in pyrite, but minor proportions may occur in other As-bearing minerals. A few base metal deposits (LaRonde in Quebec, Westmin in B.C) (Cabri et al., 1999) contain significant amounts of tennantite $((Cu,Fe)_{12}As_4S_{13})$ which is normally recovered in the Cu concentrate. The tennantite is, however, a minor carrier of the As in the ores of these deposits, but it is the major carrier of As in the Cu concentrate.
- Antimony: Antimony occurs as a constituent of tetrahedrite-friebergite-tennantite, bournonite and to a small extent in other sulfantimonide minerals.
- Bismuth: Bismuth occurs as a trace element in galena, a minor element in tennantite, and

rarely as a constituent of native Bi, Bi sulfides and Bi-Ag sulfantimonides.

- Mercury: Mercury occurs mainly as a trace element in sphalerite. Materials balances conducted by the author show that it also occurs as a trace element in another mineral (or minerals) rejected to the tailings, probably pyrite.
- Selenium and tellurium: Selenium and Tellurium occur as trace elements in chalcopyrite and galena, and as major constituents in Se- and Te- minerals. The Te in some base metal ores (LaRonde in Quebec, Westmin in B.C) occurs as a significant element in tennantite (Cabri et al., 1999)

1.2.1.2. Minerals in greisen-type tin-tungsten deposits

Greisen-type tin-tungsten deposits are processed by gravitational, magnetic and flotation techniques to recover tin, tungsten and sometimes copper. The ores are complex and have the potential for recovering a variety of other metals and minerals including molybdenum, indium, bismuth, zinc, rare earth elements, topaz, fluorite and kaolin as by-products. All the minerals that have a bearing on processing these ores need to be identified, and the compositions of some minerals, particularly the sulfides, need to be determined.

The minerals in greisen-type tin-tungsten deposits, such as the Mount Pleasant tin deposit in New Brunswick, Canada (Petruk, 1973b), occur as veins and veinlets and as disseminations in an intensely silicified rock. The silicified rock is commonly referred to as a greisen as it is enriched in topaz, fluorite and a variety of other gangue minerals including zircon, titanite and garnet. In some places the veins form stockworks and the vein and ore distributions are similar to ore distributions in porphyry copper deposits. The deposits are composed of both oxide and sulfide ores and occasionally kaolinite pipes are present.

The oxide ores contain cassiterite, wolframite, molybdenite and accessory minerals such as rutile, magnetite, hematite, ilmenite, columbite, pyrochlore and monazite. They are generally processed by gravitational techniques to recover the cassiterite and wolframite, and the tailings might be processed by flotation to recover the molybdenite. The monazite, pyrochlore, and columbite may contain trace to minor amounts of thorium and uranium.

The sulfide ores occur in zones that are enriched in either sphalerite, chalcopyrite or galena, and contain cassiterite and smaller amounts of other tin-bearing minerals which may include stannite, kesterite, Fe-kesterite, stannoidite, mawsonite and petrukite (Petruk, 1973a; Kissin and Owens, 1989). The sphalerite, chalcopyrite and pyrite may contain trace amounts of tin in solid solution. The sulfide ores also contain Bi minerals such native bismuth, bismuthinite, bismutite, and trace amounts of galenobismutite, cosalite and aikinite. The concentrations of Bi minerals may be high enough to recover the metal economically. Other minerals include tennantite, arsenopyrite, and loellingite. The indium occurs mainly as a trace element in sphalerite, kesterite, Fe-kesterite, stannite and chalcopyrite (Petruk, 1973a), but trace amounts of indium minerals such as roquesite have been found in the Mount Pleasant deposit in New Brunswick, Canada (Boorman and Abbott,1967). The sulfide ores from greisen-type tin-tungsten deposits are generally processed by gravitational techniques to recover the cassiterite and by flotation to recover the sulfide minerals.

1.2.1.3. Minerals in porphyry-copper ores

Porphyry copper ores commonly contain three ore zones, primary ore, partly oxidized ore, and oxidized ore. The primary ore generally occurs as stockworks of veins in a silicified zone of intrusive rocks, commonly granitic, and zonation of the ore minerals and the hydrothermal and metamorphosed gangue minerals is present. The primary ores are processed by flotation to produce Cu and sometimes Mo concentrates, and to recover by-products such as Au and Ag. The main copper minerals in the primary ore are chalcopyrite and bornite, although some deposits contain minor amounts of other copper minerals such as chalcocite, enargite and cubanite. Pyrite and magnetite are common gangue minerals and can be readily rejected during mineral processing. Small amounts of sphalerite and galena may be present. Many deposits contain enough molybdenite to be recovered as a separate concentrate. The gold and silver are recovered as by-products from the copper concentrate. The gold generally occurs as discrete gold grains associated with chalcopyrite, pyrite and arsenopyrite, and as invisible gold in pyrite and arsenopyrite. To obtain maximum gold recovery, it is important to determine (1) the associated minerals, (2) the percentage of gold occurring as discrete grains, and (3) the percentage of gold occurring as invisible gold in pyrite and arsenopyrite. Such data will show whether there is enough invisible gold in the pyrite and arsenopyrite to consider recovering the pyrite and arsenopyrite for the gold content.

The partly oxidized ore generally occurs above the primary ore. It contains secondary Cu minerals such as covellite, chalcocite, and digenite and some secondary hematite and goethite. The secondary copper minerals may interfere with the production of a high grade Cu concentrate by flotation, because they release Cu ions that activate the pyrite and cause it to float with chalcopyrite in the Cu concentrate. They also release sulphates which may alter the flotation conditions and reduce chalcopyrite recovery.

The Cu-bearing minerals in the oxidized zone include idaite, covellite, yarrowite, spionkopite, geerite, anilite, djurleite, chalcocite, native Cu, cuprite, malachite, brochantite, chrysocolla, and a variety of other Cu sulfates and Cu chlorides. Invisible gold that occurred in pyrite and arsenopyrite has been released, and the gold grains and inclusions that occurred as free gold may still be present as free gold. The oxidized copper ores are generally processed to recover the copper and gold by leaching. For this reason the mineralogy of the copper minerals in oxidized zones is seldom studied in detail, however, mineralogical studies can provide information about possible limitations on the copper and gold recovery by vat and/or heap leaching.

1.2.1.4. Minerals in iron ores

Most iron ores occur in iron formations that are composed of massive and disseminated iron ore minerals in siliceous rocks. Some iron formations are high grade and produce direct shipping iron ores, but most contain significant amounts of silicate minerals and need to be processed to produce iron concentrates or pellets. The main iron ore minerals are hematite and magnetite. Goethite may be present as a fine-grained dust or as botryoidal masses. Siderite, pyrite, and small amounts of ilmenite may also occur. Trace amounts of apatite and zircon are commonly present. The main silicate mineral is quartz, although a variety of other minerals including iron-rich chlorite may be present. The hematite and magnetite are usually recovered from the iron ores by gravitational and magnetic techniques, and sometimes by flotation. Flotation is commonly used as a final upgrading step (Petruk et al., 1993).

Some iron ores are enriched in manganese. The Mn-rich iron ores contain a variety of Mn oxides and Mn-bearing siderite, usually along faults. The Mn oxides include manganite, pyrolusite, psilomelane and wad, and some of the goethite and magnetite are enriched in Mn (Lewczuk, 1988). It is important to know the identities, compositions, physical characteristics and textures of the Mn minerals to evaluate whether they can be removed from the iron concentrate, and whether it is economically feasible to produce a separate Mn concentrate.

Sedimentary Minette-type iron ores occur in many parts of the world , including the Lorraine basin in France and the Peace River district in Alberta, Canada (Petruk et al., 1977). The ores generally consist of brownish, earthy friable material. The material in the Peace River iron deposit consists of oolites, siderite and earthy fragments in a matrix of ferruginous opal and clastic material. The oolites are composed of goethite, ferruginous nontronite and amorphous phosphate, and the clastic material and earthy fragments consist of illite, ferruginous nontronite and quartz. Processing the ore involves recovering the goethite and ferruginous nontronite.

1.2.2. Quantities of minerals

Information on quantities of minerals or phases in the products is needed for most applications. For example, in mineral processing, information on quantities of wanted minerals provides an indication of the value of the ore; information on quantities of gangue and other unwanted minerals helps the engineer in designing a flowsheet for rejecting minerals; and information on quantities of other minerals, such as silver-bearing minerals, provides a basis for calculating the distribution of the elements among the minerals, and helps the engineer to decide which minerals can be economically recovered.

Mineral quantities are determined by:
- image analysis of polished, thin or polished-thin sections,
- point counting of polished, thin or polished-thin sections,
- X-ray diffractometer analysis of powders,
- calculations from chemical assays,
- a combination of the above.

The easiest method of determining mineral quantities is by X-ray diffractometer analysis, but the results are generally semi-quantitative. In contrast, calculations from chemical assays produce the most accurate data for mineral quantities, but require supporting mineralogical information. Even so, the calculations cannot be performed in many instances because (a) some elements occur in several minerals and (b) element distributions among minerals cannot be defined from assay data and qualitative mineralogy.

The quantities of silicate minerals are usually determined by X-ray diffraction analysis of finely crushed material by using the ratio $I_{mineral}/I_{corundum}$ for three or more peaks in the minerals. The minimum quantity that can be determined is usually between 1 and 5 %. To improve the accuracy of analysis and to reduce the effects of preferred orientation, Szymanski and Petruk (1994) used all the major XRD peaks in the minerals in samples composed of quartz, mica and silicate minerals. Newmount Exploration routinely use XRD in their exploration program to find mineral assemblages that host Carlin-type gold deposits in Nevada (Hausen et al., 1982) .

The quantities of ore minerals (down to about 0.1 %) are usually determined by image analysis or point counting. Unfortunately, due to sample preparation, the quantity of mineral displayed on the surface of the polished section is not always representative of the sample. It is recommended that the samples be analysed chemically for at least one element prior to image analysis so that a comparison can be made between the calculated and analysed mineral quantities. If the difference is greater than 20% of the amount present, a second polished section should be prepared and analysed. The problem of preparing non-representative polished sections occurs because, during sample preparation, the particles settle at different rates in the mounting medium , depending upon their specific gravities. The heavy particles settle at the face of the polished section, whereas, many of the lighter particles are frozen in the resin before reaching the face. This has a significant effect on the distribution of minerals in polished sections. For example, The author observed that the analysed content for hematite (S.G. 5. 2) in a hematite-quartz ore was commonly 1.3 X its true amount, and the quartz content (S.G. 2.67) was 0.7 X its true amount. This problem is particularly acute with coarse-grained fractions (e.g. coarser than 48 mesh or 295 μm).

1.2.3. Size distribution

Size distribution is a basic parameter for characterizing particles and mineral grains. The size distribution is usually determined by measuring the proportion of material in each size range within a sequence of sizes. A common size sequence is the Tyler series of mesh sizes where each size is related to the next by $\sqrt{2}$, as shown in Table 1.1. In mineral processing the size distribution data are used for determining whether the particles have the proper size range for the processing equipment, for establishing mineral liberations, and for predicting the required degree of grinding to liberate minerals from an ore.

1.2.3.1. Predicting grind

One method of predicting the required grind involves measuring the size distribution of minerals in uncrushed ores and applying liberation models (King, 1979; Finch and Petruk, 1984; Petruk et al., 1986; Petruk, 1986: Lin et al., 1987). The size distribution of minerals in an uncrushed ore can be measured automatically by image analysis, or manually by point counting. Depending upon the mode of occurrence of the mineral, different analytical techniques need to be used to measure size distributions.

- if the mineral occurs as discrete grains in a matrix (Figure 1.1a), the size distribution can be determined by measuring either (1) the area covered by each grain, (2) the intercept length across each grain, or (3) counting the number of grains in each size range,
- if the mineral occurs as interconnected grains or veinlets (Figure 1.1b), the size distribution of the grains can only be determined by measuring intercept lengths.

When the area of each grain is measured, the percent of area of the grains in each size range is assumed to be equal to the quantity of mineral in each size range. When the intercept length is measured, the proportion of intercept length across grains in each size range is assumed to be equal to the quantity of mineral in each size range. When grains are counted, the number of grains in each size range is assumed to be equal to the quantity of mineral in each size range (Delesse, 1848, Rosiwal, 1898).

Table 1.1.
Tyler sequence of mesh sizes

Tyler mesh number	Aperture size (μm)	Tyler mesh number (Contd.)	Aperture size (μm) (Contd.)
5	3392	200	75
7	2400	270	53
10	1696	400	37.5
14	1200	500	26.5
18	848		18.75
25	600		13.25
36	424		9.38
48	300		6.62
65	212		4.69
100	150		3.32
150	106		2.35
			1.61

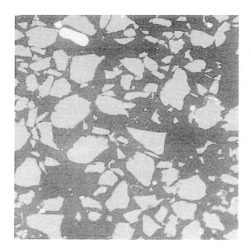

Figure 1.1a. Polished section showing discrete grains.

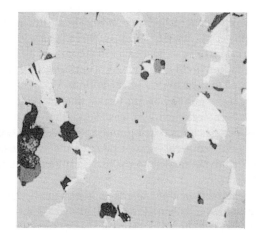

Figure 1.1b. Polished section showing interconnected galena grains in sphalerite.

1.2.3.2. Ground or powdered materials

The size distribution of particles and minerals in sands, crushed products, ground products and powdered materials is a fundamental characteristic of the product. It is used to define whether the particles are in the proper size range for the processing equipment, whether the ores have been ground sufficiently to liberate the minerals, and whether the ores have been overground. Size distributions can be determined:

- for *individual particles* by screening, cyclosizer, instruments based on sedimentation, instruments based on air dispersion, and image analysis,
- for *liberated grains in ground materials* by image analysis, or by heavy liquid separations of fractions from each size range if the mineral of interest has a much higher specific gravity than the rest of the material,
- for *unliberated grains in particles in ground materials* by image analysis. Heavy liquid separations can be used if the ore is composed of two phases, and the if the mineral of interest has a much higher specific gravity than the other material.

The size distribution of particles can be measured with an image analyser interfaced with an optical microscope using transmitted light by dispersing the particles in immersion oil on a glass slide. In contrast, if the particles are mounted in a resin and polished, different horizons of the particles are exposed during polishing and the exposed surfaces of many particles are smaller than their true size. Hence the sizes of the exposed surfaces of particles in polished sections do not represent the size distributions of the particles. However, it has been determined empirically, that when the powdered material that has been mounted in the polished section, has a wide size range of particles (>5 Tyler mesh sizes), the size distribution of the exposed surfaces of the particles in the polished section is nearly the same as the size distribution obtained by sieving the sample (Figure 1.2) (Petruk, 1978). The empirical relationship has not been validated mathematically. Nevertheless, in view of the observed empirical relationship, it is proposed that the size distribution of liberated mineral grains in a sample may be determined by image analysis. This would be achieved by:

1. analysing a polished section of each sieved fraction coarser than the finest sieve (usually 26.5 μm (500 mesh)), and measuring each particle to determine the quantity and proportion of mineral occurring as liberated grains in each sieved fraction;
2. analysing a polished section of the undersize material (-26.5 μm), which contains a range of 8 Tyler mesh sizes (1.6 to 26.5 μm), and measuring each particle to determine the size, quantity and proportion of mineral occurring as liberated grains in each size range;
3. combining the data for the sample.

It is also possible to determine the size distribution of liberated mineral grains by preparing only two polished sections, +26 μm and -26 μm, analysing each one, and combining the data. The results would not be as accurate as those obtained by analysing sieved fractions, but would be adequate for routine analysis related to processing, but not for research. It is noteworthy that sieves down to 5 μm are available, but it is not practical to routinely use the fine size sieves because:

- it takes a very long time (several hours) to produce enough material for a polished section,
- the fine sieves are fragile and very expensive to replace.

Size distribution of unliberated grains in polished sections can be determined by measuring the sizes and proportions of the exposed surfaces of unliberated grains in each size range.

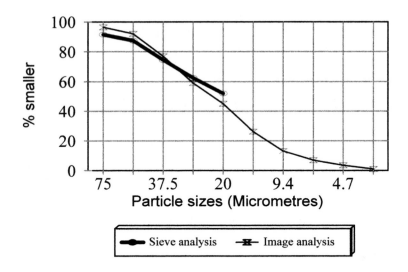

Figure 1.2. Comparison of size distributions determined by sieve analysis and by image analysis of the same material. The material analysed by image analysis was unsieved and mounted in a polished section. (Plotted as cumulative % smaller than.)

1.2.4. Mineral liberations

A primary requisite of mineralogical analysis of mill products is information on mineral liberations. A quick glance at a polished section through an optical microscope is often enough to judge whether most of the mineral grains are liberated or unliberated. This may be adequate for a general assessment, but it can lead to serious errors if used for detailed interpretations because the human eye is a very poor instrument for making quantitative measurements. Consequently, it is necessary to determine the amounts and sizes of apparently liberated and unliberated grains in each mill product by image analysis, although point counting has been used with some success (Minnis, 1984).

The degree of liberation of minerals in mill products is determined by analysing polished sections and measuring the grade of each particle (e.g. percent of mineral of interest in particle) (Figure 1.3), and the distribution of the mineral of interest among the particles (e.g. percent of mineral in each particle grade). The particles are usually classified into particle grades in incremental steps of 10 % from 0.01 % to 100 % mineral in the particle (Table 1.2). If the mineral constitutes 100 % of the particle, the particle is classified as an apparently liberated grain. If the mineral liberations are determined for particles of different sizes, and if the particle sizes have been classified by either sieve analysis or measurement by image analysis, the data can be combined into one table (Table 1.3).

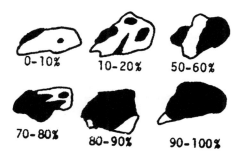

Figure 1.3. Particle grades (e.g. percent mineral of interest in particle).

Table 1.2
Example of proportion of mineral distributed among different grades of particles.

Particle grades (%)	0.1- 10	10 - 20	20 - 30	30 - 40	40 - 50	50 - 60	60 - 70	70 - 80	80 - 90	90 - 99.9	100 (free)
% mineral in particle	6.5	3.5	1.5	1	1	1	2.5	6.5	11.5	23.5	41.5

The distribution of the mineral of interest among particle grades is commonly reported as a cumulative liberation yield curve (Figure 1.4). The curve begins with the amount of mineral that is apparently liberated, and the quantities of mineral in 90 - 99.9 %, 80 - 90 %, particle categories are added sequentially.

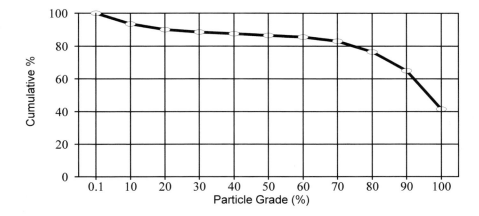

Figure 1.4. Liberation data plotted as cumulative liberation yield for mineral in particles of different grades. (Plot of data in Table 1.2)

Table 1.3
Example of presentation of liberation data in particles of different sizes

Particle Sizes (μm)	Particle Grades (%)											TOTAL
	0.1-10	10-20	20-30	30-40	40-50	50-60	60-70	70-80	80-90	90-99.9	100	
<9.4	0.07	0	0	0	0	0	0	0	0	0	0	0.11
9.4 - 13.4	0.1	0	0	0	0	0	0	0	0	0.02	0.1	0.19
13.4 - 18.7	0.1	0	0	0	0	0	0	0	0	0.02	0.26	0.41
18.7 - 26.5	0.05	0	0	0	0	0	0	0	0.1	0.15	0.65	0.94
26.5 - 37.5	0.01	0	0	0	0	0	0	0	0.1	0.34	2.37	2.84
37.5 - 53	0	0	0.1	0	0	0	0	0	0	1.98	7.67	9.83
53 - 75	0	0	0	0	0.18	0	0.1	0.1	0.15	5.07	21	26.56
75 - 106	0	0	0	0.1	0.11	0.11	0	0.14	0	13.19	26.9	40.51
106 - 150	0	0	0	0	0	0	0.2	0	0.16	6.21	11.4	17.93
150 - 212	0	0	0	0	0	0	0	0	0	0	0.68	0.68
TOTAL	0.33	0.1	0.1	0.16	0.31	0.14	0.27	0.25	0.47	26.96	70.9	100

Note 1: The total amount of mineral is 100%, and it is distributed among particles of all sizes and grades.
Note 2: The table shows that 70.93% of the mineral in the sample is liberated and 26.96% is in particles that contain 90 to 99.9 % of the mineral (bottom row).
Note 3: The table shows that 40.51% of the mineral is in particles that are 75 to 106 μm in size (last column).

The degree of mineral liberation is measured by image analysis using either an area method or a linear intercept method. The area method involves measuring the exposed area of the mineral of interest and of the host particle in polished sections, and calculating the percent of mineral in the particle (i.e. particle grade). The particle grade delineated by this type of measurement relates to volume percent of mineral in the particle, since only two phases are measured (e.g. the mineral of interest, and the rest of the particle). The particle grade can also be determined by measuring the weight % of the mineral in the particle. This involves measuring the area of every mineral in the particle, and calculating the weight percent of the mineral of interest in the particle using specific gravities of the minerals. The measurement of particle grade on the basis of the weight percent of mineral in the particle is a superior measurement, as it provides a better correlation with the actual behavior of minerals during processing (Petruk and Lastra, 1997). Currently the measurement of the area of every mineral in the particle is made with the QEM*SEM at CSIRO in Australia and with the MP-SEM-IPS at CANMET in Canada.

The linear intercept method involves measuring the length of the linear intercept across the exposed area of the mineral of interest and of the host particle in polished sections. The particle grades delineated by this type of measurement relate to volume percent of mineral in the particle.

It has been observed that the reproducibility of the measured value for apparently liberated grains (e.g. 100 % mineral in the particle) is poor, particularly when analysing minerals that are intimately intergrown in fine-grained ores. In contrast, the reproducibility of the measured value is good for minerals in particles containing > 90 % (also >80 %, >70 %, etc.) of the mineral of interest. The phenomenon of poor reproducibility of data for apparently liberated grains occurs because of particle orientation, slicing effect during polishing and the presence of very minute grains of other minerals at the edges of nearly liberated grains. In particular, particles with ~99% mineral of interest may or may not be measured as apparently liberated grains, depending upon the orientation of the particle in the polished section.

The phenomenon of poor reproducibility for apparently liberated grains is not significant from an operational standpoint because it is not economically feasible to operate concentrators to recover only particles that are totally liberated, thus concentrates always contain both liberated and partly liberated grains. Hence, it is not necessary to measure the quantity of apparently liberated mineral to define the liberation. Instead, liberation can be defined as amount of mineral in particles that contain >90 % of the mineral of interest. Such usage is desirable because it relates to the recoverable portions of the minerals. Furthermore, mineral processing engineers usually request that only one number be used to define mineral liberation and the measurement for > 90 % (also >80 %, >70 %, etc.) particles is more reproducible than the measurement for apparent liberation (e.g. totally liberated). In addition, materials balance calculations of sphalerite recoveries in high grade zinc concentrates have shown that the highest recoveries were obtained for particles containing 90 to 99.9% and 70 to 90 % sphalerite (Table 1.4). The recovery of apparently liberated sphalerite was lower due to slime loss. This indicates that recoverable sphalerite is in particles that contain more than 70 % sphalerite, and that the value for >70 % sphalerite can be used to define sphalerite liberation. Similar observations were made for galena and chalcopyrite from base metal ores, although recoverable galena from base metal ores was in >50% particles (Table 1.4).

It has been proposed that liberation be viewed from the perspective of recoverable types of particles in saleable grade concentrates (Petruk, 1990a). It follows that the maximum percentage of mineral that can be recovered from an ore that has been ground to a specific size is equal to the percentage of mineral that is in types of particles which are recoverable in the concentrate.

Table 1.4
Recoveries in concentrates from feed, of minerals in different particle grades

Particle grade	Sphalerite (>70% particles)		Galena (>50%particles)	Chalcopyrite (>70% particles)
	Faro[1]	Trout Lake[2]	Faro[1]	Trout Lake[2]
>99.9 (free)	86.9	77.7	88.1	97.7
90 - 99.9	97.5		97.1	
70 - 99.9	96.9	83.1	94.6	98.7
70 - 90	92.9		90.6	
70 - 30		55.4		92.2
70 - 20			71.2	
<30		26.7		34.1
<20			12.1	

[1] = Petruk, 1990b, [2] = Healy and Petruk, 1990

1.2.4.1. Stereological corrections

The above discussions relate to liberation data that were obtained by either two dimensional (area method) or one dimensional (linear intercept method) measurements of three dimensional irregular shaped particles whose orientations are unknown. The values for particles that were measured as totally liberated are nominally referred to as apparent liberation because the third dimension could reveal that they are unliberated. Stereological corrections should bring the apparent liberation closer to absolute liberation. King and Schneider (1993) have developed a liberation model that corrects (apparent) liberation values that were measured by the linear intercept method for different particle grades in sieved fractions. The correction is performed by using the PARGEN software. King and Schneider tested the model with different grades of particles which they produced from a sphalerite-dolomite ore by heavy liquid separations of sized fractions, and reported satisfactory correlations with the true distribution. Recently the same researchers developed stereological corrections methods that can also be applied to measurements by the area method (King and Schneider, 1997; Fandrich et al., 1998).

Many liberation models have been proposed for correcting data obtained by the area method (Klimpel, 1984; Gateau and Broussaud, 1986; Hill, 1990; Barbery, 1991; King and Schneider, 1997; Matos, 1999, Lin et al., 1987; Lin and Miller, 1986). Lin (Lin,1996; Lin et al., 1999) tested four area method liberation models, including Barbery's BOOKING software and King and Schneider's PARGEN software. He prepared synthetic binary phases composed of glass and lead borate in particles of different grades (e.g. different quantities of mineral in particles), and produced samples composed of particles that ranged from 425 to 600 μm (28 - 35 mesh) in size. One sample consisted of one grade of particles (single composition), another covered a narrow range of grades (narrow composition), still another simulated a tailing, another simulated a concentrate, and three other samples simulated other products which had a wide distribution of particle grades. Polished sections of the samples were analysed at CANMET by the **area method**, and Lin processed the data using the different liberation models (Lin, 1996; Lin et al., 1999). It was found that none of the liberation models produced the true liberation although the "corrected" liberation sets produced by the models, as well as the measured (uncorrected) data, were relatively close to the true values. The measurements for the single composition and narrow composition samples yielded values for non-existent grades (e.g. for higher and lower grade particles than were present in the sample). The BOOKING and PARGEN liberation models improved the data significantly by removing most of the values for non-existent grades (Figure 1.5a). It is noteworthy that distributions which give single or narrow particle grades are never found in mineral processing operations. For the samples with a wide distribution of particle grades, which is the normal distribution in the feed to the concentrator, there was a close correlation between the measured and true values for intermediate grades (Figure 1.5b), but the measured values for the totally liberated grade were about 10 % higher and for the totally unliberated grade about 10 % lower. The BOOKING and PARGEN liberation models improved the values for the totally liberated and totally unliberated grades but made little difference for the intermediate grades (Figure 1.5b). For the sample with the grade distribution that simulated a tailings sample (e.g. particle grades from 0.1 % to 40 %) the measured values were reasonably close to the true distribution for all particle grades, except at the 0.1 % particle grade, where the measured value was much higher than the true value. The PARGEN liberation model brought the data very close to the true data, but the BOOKING liberation model did not (Figure 1.5c). The results for the concentrate sample were surprising, as the measured value for the totally liberated grade was about 10 % lower than the true value. The liberation models made the data worse. Obviously poor data were obtained, probably by preparation of non-representative polished sections.

Fandrich et al. (1998) tested two stereological models: allocation method and kernel transformation method, on 21 grade fractions produced from an iron oxide ore composed of hematite, magnetite and gangue which was predominantly quartz. The grade fractions were produced by heavy liquid and fluidization bed separation techniques, and were analysed with the QEM*SEM by areal section liberation measurements. Fandrich et al. observed that, as in Lin's observations, the measured values for a very narrow grade fraction had data for non-existent higher and lower grade particles than were present in the sample (Figure 1.6a). The transformation method improved the data significantly by removing nearly all of the values for non-existent grades. On the other hand, for samples with the wide distribution of particle grades, the measured values were reasonably close to the true liberation but, as in Lin's data, for the totally liberated and totally unliberated particles, the measured values were 10 % lower and 10 % higher, respectively. The two liberation models corrected the measured data and produced values that were closer to true liberation than the measured data (Figure 1.6b). It is considered noteworthy that, despite careful sample preparation, Fandrich et al. (1998) also had a case where the "corrected" liberation set was worse than the measured liberation. Hence bad data cannot be corrected.

18

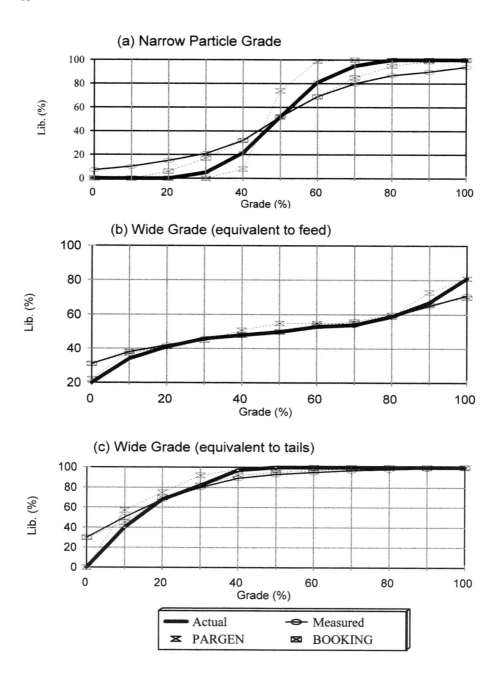

Figure 1.5. (5a, 5b and 5c). Actual, measured and corrected liberation at each particle grade (Simplified from Lin, 1996). It is noteworthy that the measurements do not necessarily relate to distributions in the samples.

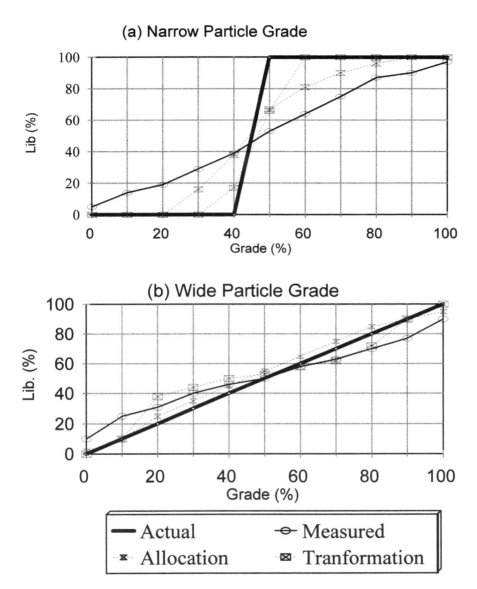

Figure 1.6. (6a and 6b). Actual, measured and corrected liberation at each particle grade(simplified from Fandrich et al. 1998). It is noteworthy that the measurements do not necessarily relate to distributions in the samples.

Lin's tests (1996) of the BOOKING and PARGEN liberation models, and the study of Fandrich et al. (1998) have shown that the liberation models:
- improve the values for totally liberated and totally unliberated grains,
- improve the grade distributions,
- produce liberation sets that are worse than the measured liberation sets when the polished sections are not representative of the sample, which is common, despite careful sample preparation.

It is considered significant that the BOOKING and PARGEN liberation models did not improve the values for intermediate particle grades in samples that had a wide distribution of particle grades. Furthermore, in some instances for samples with a wide range of particle grades, the measured (uncorrected) data for intermediate grade particles were as good as data that was corrected by the Booking or Pargen liberation models.

1.2.4.2. Measurements of mineral liberations in sieved fractions and unsieved samples

All the liberation models developed to date use sieved fractions, and provide information on liberations for minerals in each sieved fraction. Liberations for minerals in samples have to be calculated from the liberation data for sieved fractions. Unfortunately most of the particles in many flotation products are smaller than the smallest conventional sieve (26 µm or 500 mesh). It is not practical to routinely prepare sieved fractions of material that ranges from about 1.6 to 26 µm (a range of 8 Tyler mesh sizes), because it takes very long to sieve enough material for preparing polished sections of 5 - 10 and 10 - 15 µm fractions, and the sieving operation may destroy the small size (expensive) sieves. Fractions produced with a cyclosizer can be used, but the true sizes of the cyclosizer fractions are not known since they are calculated on the basis of the weight of the particle. Hence a large particle composed of a mineral with a low S.G. would be in the same fraction as a small particle composed of a mineral with a high S.G.. Furthermore it takes a long time to collect the final (very fine-grained) cyclosizer fraction (cone 5) by sedimentation.

Tests were conducted to determine whether the same liberation values would be obtained for the minerals in the sample by analysing unsieved material of the sample, as when analysing sieved fractions and combining the data from each sieved fraction into total data for the sample. Three samples of a ground volcanogenic base metal ore were sieved into +75, 53 to 75, 37.5 to 53, 26.5 to 37.5, 20 to 26.5 and minus 20 µm fractions. The liberations of sphalerite and of galena in each sieved fraction and in the unsieved sample were measured. The liberation data for the sieved fractions were calculated into liberation data for the total sample on the basis of weight percent of each sieved fraction and quantity of mineral in the fraction. The results showed that the mean apparent liberation (amount of particles with a grade of 100 %), calculated by combining the data from sieved fractions, was about 12% ±7% higher than the mean measured apparent liberation for the unsieved samples. In contrast, the measured proportion of mineral occurring in >70% particles in the combined sieved fractions was only slightly higher (4.7% ±1.5%) than the measured proportion in the unsieved sample (Figure 1.7a).

Tests were also conducted to determine whether the same liberation values would be obtained by measuring liberations of sphalerite and galena in individual sieved fractions, as by measuring the liberations of the minerals in the same sized particles in the unsieved sample. The liberations of sphalerite and of galena in 37.5 to 53 µm, 26.5 to 37.5 µm and 20 to 26.5 µm sieved fractions were measured first. The liberations in the unsieved samples were determined by: (1) measuring the sizes of the particles in the unsieved sample, (2) measuring the liberations for the sphalerite and galena in particles within the size range, which is equivalent to the size range of the sieved fractions. The Results showed that the average difference between measured values for the sieved fractions and for the unsieved samples was 38% ±11% higher for apparently liberated particles (100 % grade) in the sieved fractions than in the unsieved sample; 6.0% ±0.8% higher for particles containing >70% of the mineral; and smaller differences for particle grades of <70 %. The cumulative liberation yields that were obtained by analysing sphalerite in a sieved 26 - 37 μm fraction, and in 26 - 37 μm particles in an unsieved sample are given in Figure 1.7b as an example.

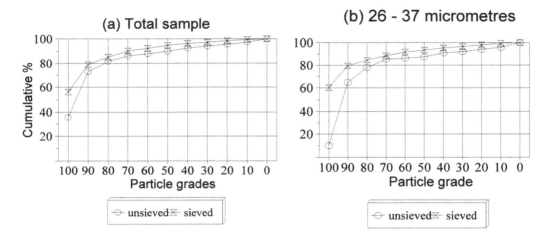

Figure 1.7a. Cumulative liberation yields for sphalerite in a Cu-Pb rougher concentrate, determined by (1) analysing sieved fractions and combining the data (sieved), and (2) analysing an unsieved sample of the same concentrate (unsieved).

Figure 1.7b. Cumulative liberation yields for sphalerite in a 26 - 37 μm fraction. Determined by analysing (1) a sieved fraction (sieved), and (2) an unsieved sample (unsieved). The sizes of the particles in the unsieved sample were measured with the IA, and the liberation data was measured for the 26 - 37 μm particles.

The above comparison shows that the measured apparent liberations (e.g. liberated particles or 100 % grade) of specific sized particles in unsieved samples are not the same as the measured apparent liberation in each sieved fraction of the same sample. In fact the difference between the observed apparent liberation (e.g. particles of 100 % grade) in sieved fractions and in particles of the same size range in unsieved samples is so large, that unsieved samples cannot be used to determine the apparent liberation of sized particles. In contrast the liberation values obtained for particles containing >70 % sphalerite by analysing unsieved samples are relatively close to liberation values obtained for the same type of particles by analysing sieved fractions. It is considered that these values are close enough for routine characterization of a process when using particles containing >70 % sphalerite as the liberation criteria.

1.2.4.3. Discussion

Liberation models are used to determine liberations of minerals in sieved fractions. Determining liberations of minerals in samples involves combining the liberation data for all sieved fractions, plus data for the portion of the sample that is too fine-grained to be sieved (usually a -26 μm unsieved fraction). Liberation data for sieved fractions can be determined by analysing the fractions and applying liberation models. Liberation data for the very fine-grained unsieved fraction can be determined by analysing the unsieved fraction and using the raw data. It has been shown that, for unsieved samples, the measured liberation values for all particles grades, except the 100% particle grade, are reasonably close to combined liberation values from sieved fractions of the same sample. For the 100% particle grade (apparently liberated), the average difference between the analysed values for the unsieved sample and value obtained by combining the data for sieved fractions is about 12% higher for the combined data from sieved fractions. Liberation models correct the analysed value for sieved fractions by around 10 %.

Petruk (1986) observed , by using materials balance calculations, that the uncorrected apparent liberation values (100% particle grade) obtained by measuring unsieved material, defined the behaviour of material in concentrators and bench tests better than corrected liberation values for unsieved samples using an empirical model that reduced the apparent liberation by 15% (Petruk, 1978; Petruk 1986).

The above observations indicate that the amount of apparently liberated mineral determined by analysing sieved fractions by the area method, should be corrected by a liberation model because the analysed amount of apparently liberated mineral is somewhat higher than true values. In contrast the analysed amount of apparently liberated mineral obtained by analysing unsieved samples by the area method seems to be close enough to the liberation of the mineral in the total sample to be used routinely in an operational setting.

It is significant that the liberation models should be applied only to data that have been obtained from polished sections that are truly representative of the sample. If there is a large sample preparation error, which may be due to particle settling, particle orientations, particle shapes, etc., or if the image analysis data are not representative of the true particle population, no stereological method has a chance of correcting the data accurately, and could make the data worse (Fandrich et al. 1998; Lin, 1996). Hence the sample preparation error, which is commonly the case in an operational setting, overrides the benefits of stereological corrections. In this case the liberation models cannot be applied. On the other hand, in an operational setting reasonable errors can be tolerated if the samples in a suite are analysed in the same manner by the same technique, and if materials balance is used to bring the liberation data to best fit (Laguitton, 1985; Hodouin and Flament, 1985; Petruk, 1988b). In this case the measured liberation data (without corrections, or corrected to best fit by materials balance) can be used to interpret mineral behaviours during mineral processing. Measured mineral liberation data have been used in this manner in the CANMET laboratories during the period of 1975 to 1999 and the interpretations that were made for mineral behaviours corresponded to observed mineral behaviours during mineral processing. In one instance uncorrected liberation differences, as small as 2 %, explained why different recoveries were obtained by processing an ore that was ground in a ball mill than in an autogenous mill (Petruk and Hughson, 1977).

1.2.5. Textures

All textures, including crystallinity, grain boundary relations, grain orientations, fractures, veinlets, etc. have a bearing on processing ores and materials, but the sizes of mineral grains and bonding between the grains are the main characteristics that influence ore breakage and mineral liberations. In particular, when ores are ground to the same particle size distribution, the liberations of mineral grains that are strongly bonded to each other will be lower than the liberations same sized mineral grains that are weakly bonded to each other. A strong bonding develops between grains during crystallization and recrystallization, and size reduction of pieces composed of strongly bonded minerals occurs by random breakage during grinding. In contrast, weakly bonded minerals tend to break along grain boundaries (preferential breakage).

Grain boundary irregularities provide a measure of the extent of intergrowth of minerals and, in turn, of bond strength between the grains. In particular, sinuous grain boundaries show strong intergrowths and probably a strong bond, whereas straight grain boundaries show no intergrowths and probably a weak bond. Grain boundary irregularities can be measured with an image analyser that has a binary thinning routine or its equivalent. The length of a grain boundary divided by the length of a median line along the grain boundary would provide a measure of the degree of intergrowth, and possibly of bond strength. Another way of detecting weakly bonded grains is by observing that polished sections of weakly bonded grains display incipient fractures and pits along grain boundaries, whereas polished sections of strongly-bonded grains do not display such fractures or pits (Petruk, 1988a).

It is interpreted that the first breakage of an ore occurs along fractures. This breakage may induce other microfractures. The next level of breakage occurs by separating mineral grains that are weakly bonded. If there is interstitial material between the matrix minerals it will fall out and become largely liberated, even if the grains of the interstitial material are much smaller than the size distribution of the ground material. Similarly if a mineral breaks along cleavage planes, the inclusions that occur along these planes may roll out. In particular, it was observed that ground sphalerite from the ore of Brunswick Mining and Smelting (BMS) contained only 10 % as much chalcopyrite inclusions/exsolutions (chalcopyrite disease) as sphalerite in unground ore. The final level of breakage is random breakage across grain boundaries.

Recrystallized volcanogenic pyrite-rich ores tend to contain relatively large recrystallized pyrite cubes which have several effects on mineral beneficiation. One effect is that the main trace elements, particularly gold and silver, have been largely expelled from the pyrite during recrystallization (Petruk and Wilson, 1993). A second effect is that the recrystallized pyrite cubes are difficult to break during grinding since they have to be broken across grain boundaries by random breakage, and increased grinding is required to reduce the grain size of the ore. In particular, a grinding test was conducted on the volcanogenic base metal ore from the Faro deposit, which has large amounts of recrystallized pyrite porphyroblasts, and on the base metal ore from Brunswick Mining and Smelting (BMS), which has very few pyrite porphyroblasts. It took 60 minutes to reduce the Faro ore from 80 % minus 1600 µm (10 mesh) to 80% minus 38 µm, and 23 minutes to reduce the BMS ore to the same size (Petruk, 1994).

Ore minerals such as silver and gold commonly occur in veinlets in the ore and rock. During grinding the veinlets separate from the rock, and may be liberated at a relatively coarse grind. As an example, the gold in volcanogenic rocks occurs in narrow veinlets and along grain boundaries in pyrite (Healy and Petruk, 1990d). During grinding the minute gold grains break away from the ore minerals and pyrite and are concentrated in the Cu concentrate. Consequently, tailings from such ores contain only encapsulated gold grains in recrystallized pyrite and invisible gold in pyrite.

1.2.6. Mineral associations

Associated minerals are minerals that occur adjacent to the mineral of interest in ores and in particles containing unliberated grains of the mineral of interest. The associated mineral may have a bearing on processing, therefore a technique for determining mineral associations is required. Four techniques have been used by the author and personnel at CANMET (Lastra and Petruk, 1994):

- binary minerals technique,
- dominant mineral in particle,
- average composition of average particle, and
- proportion of mineral that is in contact with another mineral.

1.2.6.1. Binary mineral technique

The binary mineral technique is used only with an optical microscope for manual grain counting of mill products, and has been used widely for many years. The technique, if applied directly to image analysis, does not give useful results because most particles that contain the unliberated mineral are frequently not binary particles, but are particles that contain inclusions of two or more other minerals. The manual grain counting is performed by a microscopist who makes a judgement about which mineral is the associated mineral in the so-called "binary particle" and counts it accordingly. In contrast the image analyser detects all minerals, even if they are present in very small quantities.

1.2.6.2. Dominant mineral technique

The dominant mineral technique is a variation of the binary technique, but is adapted to image analysis to provide data which is equivalent to the binary particle technique. The associated dominant mineral in the particle is defined as the mineral which, excluding the mineral of interest, accounts for more than 50 % of the remaining minerals in the particle.

1.2.6.3. Average composition of average particle

The average composition of an average particle technique is performed by producing images of the particles that contain only unliberated grains of the mineral of interest, and performing a modal analysis for the image. The results give an average composition for average particles. The most abundant mineral is the main associated mineral.

1.2.6.4. Proportion of mineral in contact with other minerals

The proportion of mineral that is in contact with another mineral is determined by measuring the interface lengths between the minerals, and calculating the relative percentages of the interface lengths. This measurement is generally made on uncrushed material to determine mineral associations in an ore, but it can also be made on mill products.

1.2.7. Surface coatings on particles

Techniques have recently been developed to measure the presence and quantities of surface coatings on mineral grains, and investigations have been conducted to help mineral dressing engineers design more efficient flowsheets. The surface coatings on particles can affect flotation, and in gold ores can contribute to a loss of gold. The effect of surface coatings on particles was studied by Kim et al. (1995) to determine why base metal concentrators that produce Cu, Pb and Zn concentrates lose some Zn to the Cu and Pb concentrates. They analysed sphalerite particles by ToF-LIMS analysis and found that some of the sphalerite particles in the Pb-Cu rougher cell were coated with slightly more lead than average sphalerite particles, and these coated sphalerite particles floated and were recovered in the Pb-Cu concentrate, rather than being rejected to the Pb-Cu tails.

Nagaraj and Brinen (1997) studied the adsorption of six sulfide collectors on chalcopyrite, chalcocite, galena, pyrite and quartz with TOF-SIMS to determine molecular information and to obtain images of the collector on the particle surfaces. They detected metal-collector complexes for the collector NBECTU on chalcopyrite, chalcocite and Cu-activated quartz; for DIBDTPI collector on galena and chalcocite; and for xanthate collector on pyrite. The adsorption of collectors on chalcopyrite was found to be via Cu-collector complex and not Fe-collector complex.

1.2.8. Search for rare minerals

A major problem in mineralogical characterization is finding small grains (particularly gold) that occur in very small concentrations (as low as 0.5 ppm). The problem is solved with SEM/image-analysis by using a rare minerals search technique. The technique involves acquiring a backscattered electron image and checking to determine whether the image contains a grain with the grey level of the mineral in question. If minerals in the image do not have the desired grey level, the sample stage is moved automatically to the adjacent field, and the next image is acquired. When the specific grey level is detected, the grain or grains having that grey level are scanned with the electron beam, and either an X-ray dot map or X-ray counts of elements is obtained for each grain to determine whether the detected grain is the desired mineral. If grain scanning with the electron beam indicates that the desired mineral is not present, the sample stage moves to the next field. On the other hand, if the desired mineral is detected, the location is recorded in a file, and the characteristics of the grain are measured. Measurements include the size of the grain, and whether the mineral is free or unliberated. If the grain is unliberated, it can be analysed to determine whether it is encapsulated or partly exposed, and to determine the particle grade. If the mineral is exposed, the edge length of the exposed grain, with respect to total edge length of the particle, can be measured. The latter measurement can be related to the availability of the mineral to leaching solutions. When the analysis is completed, each grain of the rare mineral can be interactively returned to for visual examination. The search is commonly performed overnight, and the interactive visual examination is performed the next morning. A search with an optical microscope, combined with interactive examination can also be performed. A magnification of about 800 to 1000 times is needed to detect grains about 1 μm and larger which requires about 10,000 fields of view to analyse a 1 inch by 1 inch polished section. In contrast, a magnification of about 200 times is needed to detect grains 5 μm or larger, and only about 500 fields of view are required to analyse a 1 inch by 1 inch polished section. However, most grains 5 μm or larger can easily be found by scanning manually in less time.

CHAPTER 2

INSTRUMENTS FOR PERFORMING APPLIED MINERALOGY STUDIES

2.1. INTRODUCTION

The availability of mineralogical equipment plays a major role in the extent of mineral characterization that can be performed with respect to applied mineralogy. For example, the mineralogist in a laboratory at a minesite usually has only an optical microscope, sieves, assay data, and mineral separation facilities (e.g. heavy liquid separation facilities, shaking tables, laboratory flotation cells, etc.), and is expected to provide general information on mineral characteristics that affect day-to-day operations. In contrast, the mineralogist in an applied mineralogy institute has a large array of modern equipment, and is called upon to identify detailed mineral characteristics that have a bearing on processing ores and materials, on waste utilization, on the environment, etc. The modern equipment is expensive, and generally requires highly trained operators. Therefore, very few laboratories have an array of modern mineralogical equipment, but the well equipped laboratories usually perform custom work. The amount of equipment in mineralogy laboratories at most universities is usually somewhere between the amount at a minesite and the amount in a well equipped applied mineralogy laboratory. In the past the applied mineralogists at minesites determined the mineralogical characteristics with little or no outside support, whereas now, with the ease of communication and the vastly superior data that can be obtained with the modern equipment, the applied mineralogists at minesites frequently send their samples to well equipped laboratories for specialized analyses.

An early application of instrumentation in applied mineralogy was the use of mineral separation facilities such a tables, heavy liquid separators and magnetic separators to isolate mineral grains for subsequent analyses. Optical microscopes were used to identify minerals, to observe mineral textures, and to determine mineral quantities by point counting. The development of the X-ray diffractometer (XRD) made it possible to identify many minerals with a high degree of certainty, and to qualitatively determine mineral contents in powdered materials. The development of the electron microprobe (MP) was a giant step in mineralogy, as the MP made it possible to determine the major, minor and trace element contents of minerals in polished and/or thin sections without destroying the mineral grains. The scanning electron microscope (SEM) was developed before the microprobe, but was not used to any degree in applied mineralogy because the features in the secondary electron (SE) images that were displayed by the SEM could not be used to identify the minerals. The development of the energy dispersive X-ray analyser (EDS), which enabled nearly instant identification of mineral grains in a SEM image, brought wide usage of the SEM into applied mineralogy. The period of about 1960 to 1985 was prolific for development of instruments that could be used in applied mineralogy. The instruments include the image analyser (IA), proton-induced X-ray analyser (PIXE), secondary ion mass spectrometer (SIMS), and laser ionization mass spectrometer (LIMS). The IA provided an automatic means of determining mineral quantities, mineral liberations, texture analysis, a search for rare minerals, and other parameters that can be defined by grain outlines and morphological properties of minerals and particles. The PIXE could detect smaller amounts (>~5 ppm) of trace elements in minerals than was possible with the MP (>~300ppm). The SIMS could detect even smaller amounts of trace elements(>~10ppb) than the PIXE, and is particularly useful for determining the quantities of invisible gold in pyrite and arsenopyrite. The time-of-flight laser ionization mass spectrometer (ToF-LIMS) could analyse elements that are on the surfaces of particles, and ToF-SIMS is used to analyse particle surfaces including the bonding of compounds on the particle surfaces.

Other instruments are also used for mineral identification, to observe mineral textures, and to determine trace element contents. The major ones include cathodoluminescense, which differentiates minerals that have specific luminescence properties, and infra-red, which detects the presence of OH, SO, etc. cations in minerals. An emerging technique is laser ablation, inductively coupled plasma, mass spectrometry (LA-ICP-MS). It has the capability of simultaneously analysing all elements in micro samples, at detection limits between about 0.001 and 8 ppm, depending upon the spot size of the ablation laser.

Many techniques and equipment that are used for performing mineral analyses, and are standard in many laboratories, are not discussed here because most are well known, some are discussed by Jones (1987), and some are becoming obsolete. The equipment is based on specific mineral properties that include specific gravity, magnetic properties, ultraviolet fluorescence, radioactivity, differential thermal properties, and others.

An overview of some instruments that are used in applied mineralogy is presented in this chapter to provide a "road-map" to the types of analyses that can be performed by these instruments.

2.2. OPTICAL MICROSCOPES

Hand specimens, drill core samples and large pieces of material are usually examined in ordinary non-polarized light with a binocular microscope at magnifications of about 5X to 100X, to select pieces for study, and sometimes, to make a preliminary identification of the minerals and textures.

Polished, thin and polished-thin sections of the selected pieces, and of powders, are examined with optical microscopes using transmitted and/or reflected polarized light to:
• identify the minerals,
• observe mineral textures, particle shapes and mineral associations,
• measure grains sizes using a calibrated eyepiece,
• determine mineral proportions by point counting.

The preferred optical microscope in an applied mineralogy laboratory is a combination reflected and transmitted light microscope so that polished, thin and polished-thin sections can be studied with the same microscope. In a well equipped laboratory the optical microscope is used at the early stages of a mineralogical study by examining many sections, locating areas for detailed studies with other instruments, and eliminating sections that do not need further study.

Optical microscopes have been described in many books, journal literature and specialized volumes, too numerous to list. A review of publications on transmitted and reflected light microscopy was recently published by Stanley (1998), and a review of the use of the optical microscope was presented by Criddle (1998). An overview of the features of optical microscopes used in applied mineralogy was given by Jones (1987).

The basic principles of optical microscopy, and techniques for identifying minerals in thin and polished sections, are usually taught to undergraduate geology students at universities. The students, however, need experience to be able to identify minerals on sight, and to detect various textures. The experience is needed because, although the human eye is an excellent tool for detecting slight differences in colours and shades which differentiate the minerals, it takes considerable experience before the eye can detect the subtle differences between properties of different minerals in polarized light.

A laboratory at a minesite commonly has an optical microscope. The microscopist at such a laboratory usually becomes familiar with the optical properties of most of the minerals in the ore suite from the specific mine in a comparatively short period time, and uses point counting to determine mineral quantities and possibly mineral liberations. To perform the point counting the microscopist has to set-up a grid and count the mineral at each intersection in the grid. This is usually done with the aid of a point counter which has a step-scan assembly mounted on the microscope stage. The step-scan assembly advances one step on the grid each time that the identity of the mineral under the cross-hair is recorded.

2.3. X-RAY DIFFRACTOMETER

Every mineral has a unique X-ray diffraction (XRD) pattern that is dependent upon the crystal structure, and to a minor degree upon the composition of the mineral. The XRD patterns are obtained by X-ray diffraction, and are used to identify the minerals and to determine mineral quantities. The principles and application of X-ray diffraction in applied mineralogy have been briefly reviewed by Jones (1987), but a few techniques and current applications are discussed here.

Minerals are identified by using a powder camera, Guinier camera, X-ray diffractometry, and a microbeam diffractometer. The powder camera technique involves mounting a very small amount of powder on the end of a glass fibre which has been dipped in grease to make it sticky. The loaded glass fibre is secured in the middle of the camera and rotated during analysis. An X-ray pattern is diffracted onto a film which has been mounted in the camera. The positions of the XRD lines on the film are commonly read manually to identify the minerals, although computer techniques for reading X-ray films have been developed.

The Guinier camera is designed to identify minerals that diffract low-angle lines (e.g. clay minerals). The sample is prepared for analysis by grinding the material in acetone to produce a fine powder, and mounted in the sample holder by painting the slurry of acetone and powder onto a sticky tape. The sample holder is mounted in the Guinier camera, and XRD lines are diffracted onto a film. The resulting XRD pattern is read manually.

X-ray diffractometry involves grinding the powder to at least minus 325 mesh (<44 μm) and mounting the ground material as either a thin film on a sticky surface on a glass slide, or as a compacted powder in a cavity in a sample holder. The ground material on the glass slide is used when only a small amount of sample is available and only mineral identities are required. The compacted powder in a sample holder is used when the sample is analysed for mineral quantities as well as for mineral identities. The mounted sample is placed in the path of the X-ray beam so that X-rays can be diffracted by the different minerals. The diffracted X-ray signal is collected by a detector (e.g. scintillation counter). The detector sweeps in an arc across the positions of the lines diffracted by the minerals in the sample and measures the intensities of the diffracted X-rays at the different peak positions. The data can be read manually from a strip chart or recorded by computer. In computerized XRD units the minerals are identified automatically using a software package that employs search-match techniques.

A micro-beam diffractometer attachment is used to identify individual mineral grains in thin sections or small polished sections. It is operated by focussing the X-ray beam on a grain in the section and obtaining an XRD pattern.

Mineral quantities can be determined with an X-ray diffractometer, and several techniques have been used. They include:
- internal standard (Klug, 1953),
- spiking successive sub-samples with the identified mineral (Alexander and Klug, 1948),
- comparing peak intensities to intensities of pure mineral (Petruk, 1964),
- relative intensity ratios (I/I_c) (Hubbard and Snyder, 1988),
- analysing quartz collected on filters (Knight et al., 1974),
- Rietveld method of minimizing weight differences between observed and calculated intensities (Bish and Post, 1993; Mandile and Johnson, 1998).

The method of using relative intensity ratios (I/I_c) has proven to be reliable and is widely used. It is based on the relative intensities between the XRD patterns of the minerals analysed and the XRD pattern of corundum. The technique requires a library of relative intensity ratios between the minerals and corundum, but established ratios are transferrable between XRD units. Internal standards are not required for analysis, as all peak intensities are transformed to a common denominator (e.g. the peak intensity of corundum). All the minerals in the sample need to be identified and the results are normalized to 100 %. Initially only the strongest lines were compared, subsequently the 3 strongest lines were used, and in 1994 a technique was developed at CANMET to use the entire XRD pattern. The entire XRD pattern for each mineral provided a better comparison and minimized preferred orientation (Szymanski and Petruk, 1994). Preferred orientation was further reduced by using a stainless steel randomiser punch (Peters, 1970). Szymanski and Petruk (1994) used a Fein-Marquart search-match package to automatically obtain a semi-quantitative estimate of the minerals identified, and used the I/I_c ratios for all the minerals in the sample to convert the semi-quantitative estimates into quantitative results, with a high degree of accuracy. It is considered noteworthy that Hausen (1979) has routinely used the randomiser punch for many years to perform quantitative mineralogical analyses by XRD.

2.4. SCANNING ELECTRON MICROSCOPE (SEM) WITH ENERGY DISPERSIVE X-RAY ANALYSER (EDS)

The scanning electron microscope (SEM), interfaced with an energy dispersive X-ray analyser (EDS), is a microbeam instrument that is used in applied mineralogy to analyse polished and/or thin sections, as well as unmounted pieces of material. The SEM is used to identify minerals, to obtain photomicrographs that show the sizes and relationships of mineral grains, and to obtain X-ray images that show the distributions of elements in minerals. Details of the SEM are described in numerous reports in the literature. The techniques for performing SEM analyses are taught by manufacturers, and at short courses conducted by various institutes.

The SEM produces an electron beam under high vacuum. The electron beam is either scanned over the entire sample, or is focussed on a grain in the sample. The sample needs to be coated (usually a thin layer of carbon or gold) to prevent charging on the sample surface. The irradiated material in the sample produces backscattered electrons (BSE), secondary electrons (SE), X-rays and other signals. The SEM is generally equipped with BSE, SE and EDS detectors to detect these signals.

The BSE detector displays the BSE signal on the CRT screen as a grey level image that shows the distribution of the minerals in the polished or thin section (Figure 2.1). The minerals in the BSE image appear as different shades of grey, depending upon the average atomic number. The average atomic numbers of some minerals are listed in Table 2.1. Most silicate minerals have low average atomic numbers and appear dark grey in BSE images. In contrast, ore minerals have higher average atomic numbers and appear in shades of light grey to white. The shade of grey can, however, be changed by changing the contrast and brightness settings on the BSE detector. The differences in the shades of grey between the minerals can be either enhanced or reduced by changing the contrast, brightness, voltage and current on the SEM.

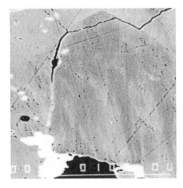

Figure 2.1. BSE image of a massive sulfide ore. It shows silicate minerals (black), pyrite (grey), sphalerite (light grey) and galena (white). Note: the gain and brightness on the BSE detector were set to bring out the silicates as black. The horizontal line in the photomicrograph is 100 μm long.

Figure 2.2. BSE image of pyrite with trace amounts of Sn. The darker grey areas are pyrite with negligible Sn contents, the lighter zones are pyrite enriched in Sn, the white Areas are galena, and the black is silicates. The horizontal line in the photomicrograph is 100 μm long.

Table 2.1
Average atomic number of selected minerals, in ascending order

MINERAL	Z value	MINERAL	Z value	MINERAL	Z value
Coal macerals	3 to 5	Hypersthene	12.9 ± 0.7	Tenorite	24.8
Diamond	5	Pyroxene	13.1	Zircon	24.8
Graphite	5	Anhydrite	13.4	Bornite	25.3
Magnesite	8.9	Biotite	13.6 ± 2.2	Sphalerite	25.4
Brucite	9.4	Wollastonite	13.6	Enargite	25.5
Lepidolite	9.8 ± 0.3	Apatite	14.1	Iron	26.0
Gibbsite	9.5	Epidote	14.2 ± 2.1	Tennantite	26.4
Bauxite	9.8	Garnet	14.3 ± 1.3	Chalcocite	26.4
Chrysoberyl	9.8	Olivine	14.6 ± 4.1	Cuprite	26.7
Beryl	10.2	Fluorite	14.6	Arsenopyrite	27.3
Kaolinite	10.2	Halite	14.6	Cobaltite	27.6
Montmorillonite	10.4 ± 0.1	Sphene	14.7	Gersdorffite	27.9
Topaz	10.5	Rutile	16.4	Stannite	30.5
Talc	10.5	Siderite	16.5	Tetrahedrite	32.5
Serpentine	10.6 ± 0.4	Tourmaline	17.8	Monazite	37.3
Spinel	10.6	Goethite	17.5	Proustite	38.9
Forsterite	10.6	Pyrolusite	18.7	Stibnite	41.1
Corundum	10.6	Ilmenite	19.0	Cassiterite	41.1
Enstatite	10.6	Psilomelane	19.2 ± 0.5	Pyrargyrite	42.4
Kyanite	10.7	Malachite	19.9	Argentite	43.0
Andalusite	10.7	Chromite	19.9	Silver	47.0
Albite	10.7	Brochantite	20.3	Wolframite	51.2
Quartz	10.8	Hematite	20.6	Scheelite	51.8
Dolomite	10.9	Pyrite	20.7	Pyrochlore	52.9 ± 1.7
Sodalite	11.1	Magnetite	21.0	Bournonite	54.4
Muscovite	11.3	Pyrrhotite	22.4	Tantalite	54.8 ± 0.9
Amphibole	11.4 ± 0.8	Linnaeite	22.4	Sperrylite	58.5
Orthoclase	11.8	Pentlandite	23.4 ± 0.1	Boulangerite	61.5
Anorthite	12.0	Chalcopyrite	23.5	Thorite	67.2
Gypsum	12.1	Millerite	23.8	Bismuthinite	70.5
Calcite	12.6	Xenotime	24.2	Galena	73.2
Ankerite	12.7 ± 0.2	Covelline	24.6	Gold	79.0

Summarized from Jones (1987).

The BSE image is the most informative image for mineralogical applications because it shows the distributions of minerals, and can provide qualitative information on trace elements in the minerals. Furthermore, when using polished surfaces, the image is free of the effects of topography, fine scratches generally do not appear in the image, and holes in polished sections are readily detected. On the other hand, relativity high voltage (>~15 kV) and current (>~15 nA) are required to produce good BSE images. An optimized BSE image is sensitive enough to display very small changes in average atomic number of the mineral, and this property is useful to show the distributions of trace elements in a mineral. For example, Figure 2.2 shows the distribution of trace amounts of Sn in pyrite from the Kidd Creek volcanogenic base metal deposit. The pyrite has zones that contain up to 0.4 wt % Sn (Cabri et al.,1985). Zones of high Sn contents in pyrite are lighter grey than zones that contain less Sn or no Sn because Sn has a higher atomic number than the Fe and S. This causes the average atomic number to be higher in Sn-rich zones than in Sn-poor zones.

The SE detector displays the SE signal on the CRT screen as a grey level SE image. The image shows the distribution of minerals and some topographic effects, as the SE signal is based on a combination of the average atomic number of the mineral and the topography of the sample. The SE image is not as useful as the BSE image for showing mineral distributions, but displays details of surface irregularities much better, and is particularly useful for observing individual particles mounted on double-sided sticky tape. The SE image can be produced at a much lower current and voltage than is required for the BSE image.

X-ray signals are detected with the EDS detector. The minerals are identified by focussing the electron beam on a specific particle and collecting the X-ray signal. In layman's terms, the EDS detector sends the X-ray signal to the EDS analyser that sorts the signal into the different elements present in the particle, and into X-ray counts for each element. The X-ray counts are recorded, and displayed as peaks on a CRT screen. An experienced microscopist can identify the mineral by glancing at the CRT screen to observe the presence and approximate proportions of elements in the mineral. If a more accurate analysis is required the EDS analyser can be programmed to perform either semi-quantitative or quantitative analysis if the X-ray signal is obtained from a smooth flat surface. In contrast, X-ray signals obtained from irregular surfaces of unmounted pieces or particles are not good enough for even qualitative analysis of the element contents because of interference from the rough sample surfaces. On the other hand, it may be possible to reduce the interference when analysing small particles by changing the working distance. X-ray signals obtained from rough surfaces are, however, good enough to identify the elements in the minerals.

The standard Si EDS detector can detect elements that are heavier than sodium (atomic number 11). The Si light element EDS detector can detect elements heavier than boron (atomic number 5) which includes carbon (atomic number 6) and oxygen (atomic number 8). The Ge light element detector can detect elements heavier than lithium (atomic number 3), although some manufacturers have begun using a thicker polymer window and this has reduced the detection of light elements.

2.4.1. Low-vacuum SEM

A low-vacuum SEM was developed in Australia for mineralogical applications (Robinson and Nickel, 1979), and is now available commercially. The instrument is known commercially under various names, such as low-vacuum SEM, variable pressure SEM and Natural SEM. Robinson (1998) describes the operation of the low-vacuum SEM in the following manner. *"The electron column and electron gun are maintained at the normal high vacuum ..., while the specimen chamber is maintained at a low vacuum .... The air in the sample chamber ... is ionized by the primary electron beam, conducting electricity sufficiently to allow the electrons absorbed by the sample to leak through the air to a ground contact, so that no coating is needed, even at high accelerating voltages (Moncrieff et al., 1978)"*. The low-vacuum SEM has a great advantage because the samples do not need to be coated for analysis at low vacuum, hence many samples can be scanned as quickly as with an optical microscope. Robinson (1998) also states *"The magnification range and depth of field of the SEM substantially exceed those of the optical microscope. Surface features on rough grains can be imaged with a level of detail not possible with an optical microscope"*. Furthermore, wet samples can be analysed so that a slurry from a flotation operation can be analysed within a very short time after it is collected. The low-vacuum SEM can be switched to high vacuum for analysis of minerals that require high voltage and current. On the other hand the low-vacuum can be used with a high enough voltage and current to get good BSE images. In fact the BSE images produced by the low-vacuum SEM At CSIRO in Australia were found to be ideal for mineralogical studies (Robinson, 1998).

Robinson (1998) pointed out that the low-vacuum SEM is as fast and as easy to use as an optical microscope for an experienced mineralogist, but the SEM provides more information. Similarly the author found that the conventional high-vacuum SEM is as easy to use as an optical microscope, but much more information can be collected in the same time period. In particular the sample can be scanned as quickly as with an optical microscope, especially if a real time BSE detector is used. In addition, nearly all the minerals can be immediately identified with the EDS, and the presence of minor elements in the minerals can be readily detected. On the other hand minerals that have the same composition, such as pyrite and marcasite, cannot be differentiated with a SEM, but are readily differentiated with an optical microscope. Both the low-vacuum SEM and the conventional high-vacuum SEM, therefore, provide a convenient compliment to the optical microscope. The conventional high-vacuum SEM, however, requires a coating, usually carbon, on the polished surface of the sample. If the sample needs to be re-observed under the optical microscope, the coating needs to be removed. A coating is not required when using the low-vacuum SEM and the sample can be easily re-examined by optical microscopy. The optical microscope, conventional high-vacuum SEM and low-vacuum SEM are used routinely in the Process Mineralogy Laboratory at CANMET.

2.5. ELECTRON MICROPROBE (MP)

The electron microprobe (MP) has played a major role in mineralogy, since it was developed in the late 1950's. It is used to determine the compositions of minerals in polished sections without destroying the mineral grains. Like the SEM, it is a microbeam instrument, but wavelength spectrometers (WDS) are used to detect the X-ray counts from the sample surface instead of, or in addition to, the EDS. The WDS are set at specific positions to detect the X-ray counts for specific elements. Unlike the EDS, which detects and counts the X-rays signals for all elements at the same time, the WDS counts X-rays signals for only one element at a time. The WDS can count many more X-rays for the specific element in the same length of time and hence is more accurate than EDS, and has a lower detection limit. Details of the microprobe are described in numerous reports in the literature, and techniques for performing microprobe analyses are taught by manufacturers, and at short courses given by various institutes.

The electron microprobe is generally equipped with SE and BSE and EDS detectors for easy viewing of the sample and for fast mineral identification. The MP is used to analyse grains, as small as 5-10 μm, for minor and trace elements with the WDS, and for major elements with EDS. The major element contents are commonly determined with EDS because the EDS analysis is accurate enough for major element analysis, and the EDS analysis is faster than WDS if the mineral contains many major elements. If, however, the mineral contains only a few major elements it may be just as fast to determine the major element contents with the WDS.

The analysis is usually performed by writing a macro which would:
- control the spectrometers to move to the peak positions of the elements to be analysed,
- Set the count time for each peak (commonly 10 seconds or a maximum number of counts for major minerals, and up to 100 seconds for minor or trace minerals) (minimum detection of trace elements is 200 to 500 ppm, depending on mineral),
- insert beam blanking at appropriate times,
- collect data from the standard under the established analytical conditions,
- move sample to first point to analysed,
- collect data for unknown under established analytical conditions,
- move the sample to the next point to be analysed.

Since the analyses are done in an unattended mode in modern microprobes, the set-up includes setting the coordinates for the points to be analysed. Many points can be analysed in a relatively short period of time by this technique. A wide variety of many silicate and ore minerals have been analysed at CANMET for many years. Recently there has been a demand for determining trace elements in diamond-indicator minerals such as spinel, ilmenite, garnet, etc. A routine program was established to analyse either 9 or 13 elements (as requested by the customer) at each point in the indicator mineral. A method was developed to mount many particles, at known locations in a polished section, and to analyse about 50 to 1,000 selected points in the particles by running the program overnight or over a week-end.

Recently an analytical technique, that can detect trace elements in the 5 to 20 ppm range, was developed for modern electron-microprobes (Robinson et al., 1998; Cousens et al., 1997). The technique uses a high accelerating voltage, a high probe current, long counting times, and background points that are near the peak and without interference. Kojonen and Johanson (1999) used the technique to determine the amounts of invisible gold in zoned pyrite in the Suurikuusikko Au deposit in Central Lapland. They determined that the average gold content in pyrite was 46 ppm with a range from below the detection limit of 22 ppm to 585 ppm, and the average gold content in arsenopyrite was 267 ppm with a range from below 22 ppm to 964 ppm. The measuring time was 600 seconds on both the peak and background at 35 kV accelerating voltage and 490 nA probe current.

The modern microprobe also has element mapping facilities, whereby different concentrations of elements are shown in different colours. This technique is useful to show the distributions of minerals that have different quantities of the same element (Figure 2.3). Unfortunately, it takes a long time to produce the maps, although the quality of the map increases as the mapping time increases.

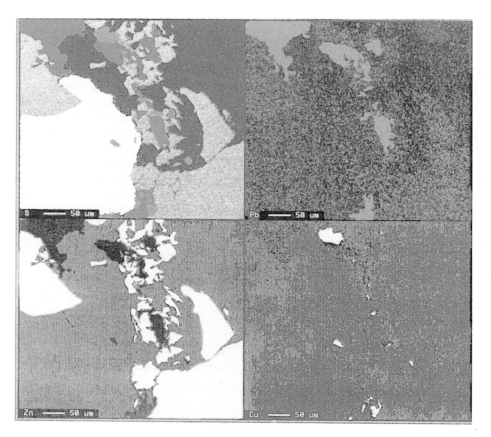

Figure 2.3. X-ray map converted to black and white from a colour image produced with a JOEL 890 microprobe by scanning an ore sample for 3 hours. It shows the concentrations of S, Pb, Zn and Cu, with grey level representing concentrations. White = high, mottled grey = medium, dark grey = low. The distributions of pyrite (white), sphalerite (mottled grey) and galena (dark grey) are displayed in the S map. Similarly the distributions of galena, sphalerite and chalcopyrite are displayed in the Pb, Zn and Cu maps, respectively.

2.6. PROTON INDUCED X-RAY EMISSION (PIXE)

The PIXE is a microbeam analytical instrument that is used for multi-element quantitative analysis of trace and major elements in selected minerals in polished or thin sections. Cabri and Campbell (1998) reported that in most cases the analyses are performed for elements that range from Fe to U (atomic numbers ≥ 26) at levels of detection commonly in the range of a few ppm.

The PIXE analysis is non-destructive, but according to Cabri and Campbell (1998), the penetration of protons generated by the PIXE is much deeper than the penetration of electrons generated by the MP, and X-rays are produced from well below the surface of the mineral. A large surface area is required for analysis, ideally 80 μm in diameter, although grains as small as 50 μm can be analysed. The PIXE analysis is similar to the electron microprobe analysis but there are several differences. The main one is that the signal to noise ratio is better in the PIXE than in the MP, hence lower detection limits are obtained in the PIXE. Another difference is that the X-rays produced by the MeV-energy protons require less corrections to yield quantitative analysis in comparison to X-rays produced by the KeV energy electrons in the microprobe (Cabri and Campbell, 1998). Cousens et al. (1997) concluded that the accuracy of the PIXE and MP are comparable, but that trace element analyses with a MP require considerable attention to choice of background position and correction for overlapping peaks. Cousens et al. (1997) also stated that the PIXE *"generally has the advantages of a large number of X-ray lines and trace element detection levels being generally smaller by a factor of two"*. Cabri and Campbell (1998) summarized many examples of PIXE analyses in ore mineralogy studies and listed trace elements that could be usefully analysed for in base metal ores, Ni-Cu sulphide ores, porphyry copper ores, sulfide Au ores, and in diamond- and kimberlite-indicator minerals.

The PIXE is an expensive instrument, hence only a few laboratories in the world have the equipment, but most provide custom analyses. The samples submitted for custom analysis need to be documented and well-prepared. The documentation is needed because analysis for trace elements involves standardization that requires knowledge of major element concentrations. Ideally the documentation would include:
- results of a microscopical study of the sample,
- microprobe analyses for the significant minerals in the sample,
- a photographic record that clearly outlines the grains for PIXE analysis.

2.7. SECONDARY ION MASS SPECTROMETER (SIMS) AND (ToF-SIMS)

The secondary ion mass spectrometer (SIMS) can produce analytical information that cannot be obtained by other microanalytical methods. It can detect and measure all the elements in the periodic table in concentrations from ~1 % to ~10 ppb (McMahon and Cabri, 1998), and has the ability to discriminate between the isotopes of many elements. The element and isotope contents can be measured with a sensitivity that is as high as the sensitivity of bulk analysis methods, such as atomic absorption (McMahon and Cabri, 1998). The SIMS can also provide molecular information about organic species on the sample surface, and is capable of providing an image that shows the distribution of the species.

The high sensitivity of SIMS makes it an excellent tool for the analysis of trace elements in minerals. It has been widely used in applied mineralogy to measure quantities of invisible gold (>0.1 ppm) in the minerals (especially in pyrite and arsenopyrite). The technique has also been used in many geochemical, environmental and geological studies. One geochemical application is the use of rare earth elements (REE) as petrogenic indicators. The REE usually occur in the minerals at concentrations below the detection limit of the MP, but can be readily detected and analysed for with SIMS (McMahon and Cabri, 1998). Nesbitt and Muir (1998) described an environmental application of SIMS where they analysed weathered plagioclase to study processes affecting dissolved Al and Si in acidic soil solutions. Common geological applications of SIMS are isotope analyses, especially of Pb, Os, Hf, and Sr (Stern, 1998).

SIMS analyses are performed on individual grains, larger than 40 to 50 μm, mounted in polished sections. The grains are usually pre-identified, and ideally pre-analysed with a MP. The SIMS analysis is performed by sequentially removing the grain surface in a small irradiated area with an incident primary ion beam by a process termed "sputtering" (McMahon and Cabri, 1998). A secondary ion is produced and the secondary ion intensity of the element or isotope of interest is measured in steps as a sputter crater is created in the mineral grain. The depth of the sputter crater is measured with a profilmeter, and the sputter time is converted to depth.

Quantitative analyses require standardization, as the secondary ion yields vary by several orders of magnitude and are very sensitive to chemical changes at the sputtered surface. Because of these effects, mineral matrix-matched standards are required for quantification. An ion implantation technique has been developed to prepare ion-implanted standards (Leta and Morrison, 1980). Once a standard has been ion-implanted and verified, quantification can be carried out on the unknown specimen by the method of relative sensitivity factors (RSF) (Wilson et al., 1989). The RSF is a calibration factor used to convert the experimentally measured ion count rate to a concentration. The RSF is unique to the sample matrix, analysed species, operating conditions and choice of calibrating species. A typical protocol for analysis would be to analyse the standard twice at the beginning of the day, then without changing any experimental conditions, analyse 25 or more unknowns and re-analyse the standard at the end of the day. This procedure generates four RSF values, the average can be used for the quantification of the individual grains (Cabri and McMahon, 1995).

The complexity of mineralogical samples makes it imperative to analyse as many grains of the same mineral as possible to get average elemental contents. The larger the number of analyses the more accurate the estimate of the element content and of the element distribution in the mineral.

Time of Flight (ToF-SIMS) analysis is performed to analyse the surface layers of particles. The particle is rastered using the total negative ion image as briefly as possible to minimize erosion (sputtering) of the topmost surface layer by the primary beam. The technique is particularly useful for analysing flotation collectors on particles. Nagaraj and Brinen (1997) used a Time of Flight (ToF- SIMS) spectrometer to analyse adsorbed collector species on mineral surfaces.

2.8. LASER IONIZATION MASS SPECTROMETER (LIMS) AND ToF-LIMS

The LIMS is an instrument that is capable of analysing elements and adsorbed organic compounds on the surfaces of particles, and the elements in the particles themselves. LIMS analysis has been applied to mineral processing by identifying, and determining quantities of, elements and compounds adsorbed on the surfaces of mineral particles. The particles are prepared for analysis by hand picking essentially liberated grains, about 20 to 100 μm in size, under a stereoscope with very fine-tipped tweezers and placing the particles in rows on indium foil. Indium foil is used as a substrate because it is conductive and has excellent adhesive qualities. The assumed identity of each particle is recorded, and photographic maps of the mounted particles are prepared to facilitate location of the particles while in the LIMS. The LIMS has a relatively small beam size, and mineral grains as small as 5-10 μm across can be analysed (Chryssoulis et al., 1992)

Chryssoulis et al. (1994) described the LIMA 2 model LIMS manufactured by Cambridge Mass Spectrometry and retrofitted with a second laser as follows: *"The first laser runs perpendicular to the sample and its function is to remove or "ablate" material from the surface (it is focussed on the particle surface with a visible coaxial He-Ne beam). The second laser, coupled to the ablator with delay times in the range of 700 to 1400 nanoseconds, runs parallel to the sample about 600 microns above the surface. Its function is to ionize the ablated neutrals. Both the distance from the surface and delay time are adjustable to maximize the volume of interaction of the plasma produced by the lasers. ....surface sensitivity ... is increased by the ability to control the ablation power such that just enough material is made available for post ablation ionization (PAL). For analysing negative ions (O^-, OH^-, etc.) only the ablation laser is used (LAO mode)......... By firing the ablator laser repeatedly on the same spot depth profiles can be obtained.*

The ions produced are extracted .....and focussed into .... a Time-of-Flight (ToF) drift tube. In the drift tube the ions are separated according to the mass overcharge ratio with the light ions arriving first in the electron multiplier. Time resolved mass spectra are transferred .... to a computer for storage and further processing".

LIMS data are presented as either count ratios or mass spectra. The count ratios are calculated by dividing the number of counts of the selected ion by the total number of ion counts. The mass spectra are used to qualitatively compare the surface and subsurface regions. In contrast count ratios provide a simple method of comparing data and are more precise than absolute counts, which are affected by changes in operating conditions and particle geometry.

Depth profiles define the mineral, detect inclusions in the mineral, and determine the relationship between the mineral and surface layer. Repetitive signals with depth indicate that the underlying material is monomineralic, a new signal indicates an inclusion, signals which decrease or disappear altogether in a second pulse on the same spot are indicative of a true surface layer.

The LIMS is available at very few laboratories, but has been widely used to determine the quantities of elements and of adsorbed organic compounds on the surfaces of mineral particles.

2.9. CATHODOLUMINESCENCE (CL)

Cathodoluminescence is a phenomenon whereby the electrons in a mineral are excited by an incident energy source, and the decay of the excited electrons to a ground state emits some or all of the excitation energy as a visible spectrum of various colours. The colours of the CL minerals are particularly useful for detecting the presence, sizes and textures of minerals and mineral grains that cannot be detected and/or observed by other methods. Hayward (1998) presented a review of CL and its application in the mining industry, and (Hagni, 1985, 1986) published several applications of CL to mineral processing and ore genesis.

According to Hayward (1998) the incident energy source is supplied by a CL system that can be mounted on either an optical microscope or a SEM. The optical and SEM systems perform many of the same tasks, but have different non-overlapping capabilities. The optical microscope CL systems are commercially available as cold cathode and hot cathode systems. The cold cathode systems operate at beam voltages of up to 30 kV and 2 mA, and can take samples up to 17 mm thick and 50 X 70 mm across in the vacuum chamber. They have the advantage of being relatively cheap and rugged. Furthermore they permit rapid switching between observations in CL and in plane- or cross-polarized transmitted or reflected light, and it is not necessary to carbon coat the samples. Some systems have EDS capabilities which can combine compositional data with optical observations.

The hot cathode systems have high beam powers and stability, as well as the visual imagery of a petrological microscope. The electron gun directs the focussed beam upwards onto an inverted thin section that has been carbon coated. Thick samples cannot be studied in hot cathode systems. One advantage of the hot cathode system is the short working distance so that lenses of higher numerical apertures can be used.

Almost any petrological microscope with sufficient clearance between the stage and objective to accommodate the vacuum chamber (at least 5 cm) can be used for optical CL (cold cathode and hot cathode). All the accessories unnecessary for CL should be removed from the optic path in the microscope, as light losses within the optical system can have a radical effect on the level of details that are visible.

The SEM CL detector can be mounted on most SEMs and does not interfere with backscattered (BSE) secondary (SE) and X-ray detectors (EDS). The SEM CL system has greater resolution than the optical CL, and observations and spectroscopic measurement are possible from areas ≥ 1 μm^2. An advantage of the SEM CL is that the BSE and SE images can be observed, and the composition of the mineral can be determined with the EDS. The sample has to be carbon coated for SEM CL analysis.

Cathodoluminescence occurs because of defects in the mineral structure. There are many types of defects, such as vacancies, non-stoichiometry, substitutional chemical impurities, disorder, lattice damage, dislocations and structural defects. The defects within the mineral structure that influence the emission of CL are referred to as *centers*. Those centers that promote luminescence are called *activators*, whereas those that inhibit luminescence are known as *quenchers*. Other centers appear to facilitate luminescence emitted by activators without themselves emitting radiation (*co-activators*). The fundamental causes of CL are complex and are discussed in some detail by Hayward (1998). The luminescent colour of a given mineral is complex and may vary from one environment to another. For these reasons the investigator must conduct an examination to determine the identity of each cathodoluminescent mineral. Despite these variations, the colour and intensity of many minerals are so characteristic and constant for a geological environment that CL microscopy can be used for identification of minerals from specific geological environments. On the other hand, because of the variations in colour, CL is an excellent tool for identifying the depositional histories of minerals. Hagni (1985) reviewed methods of identifying and defining the mode of occurrence of collophane, a phosphorous-bearing carbonate-fluorapatite, in sedimentary iron ores. The collophane could not be studied with a reflected light microscope in polished sections because it could not be differentiated from quartz, and could not be studied in transmitted light in thin sections because it contained so much very fine-grained disseminated hematite that transmitted light would not pass through the mineral grains.

Most metallic minerals do not cathodoluminesce, but many transparent to translucent minerals do. Hagni (1985,1986) discussed and listed the CL colours of 38 minerals. Hayward (1998) listed the colours of 31 poorly defined CL minerals, and discussed the colours and the reasons for CL in 18 others. Table 2.2 lists the CL minerals and their CL colours, as reported by Hagni (1985, 1986) and Hayward (1998).

Table 2.2
Cathodoluminescent minerals and their CL colours

Minerals	Cathodoluminescent colours
Anatase	Blue
Andalusite	Dull light green, red, blue, white, none
Anglesite	white, none
Anhydrite	tan, green, light orange
Apatite	white to light purple, light blue to bluish violet, light yellow
Armstrongite	blue
Baddylite	green, orange
Barite	green, red, brown, pale violet, none
Calcite	orange, blue, violet, white, green, none
Cassiterite	yellow, orange
Cerussite	white
Chalcedony	deep purple
Clinoamphibole	bright yellowish-white,
Collophane (carbonate-fluorapatite)	bright blue to light violet, light bluish white to white, dull purple, dull brown.
Corundum	red, blue,
Diamond	bright blue, pale blue, apricot, orange, red, yellowish green, bright green, yellow, white,
Dickite	bright light blue
Diopside	dark green,
Dolomite	orange red to red, dark brown, yellow orange, yellow-green

Table 2.1. (Cont'd) Cathodoluminescent minerals and their CL colours

Minerals	Cathodoluminescent colours
Elpidite	green, none
Emerald	crimson, none
Epidote	green
Feldspar	blue, green, yellow, greenish blue, red, pink, purple, none
Fluorite	blue, blue-grey, lilac, greyish white
Garnet	dull green
Gittinsite	bright orange, orange
Halite	dull greyish-blue
Hematite	red
Hemimorphite	dull green to greenish blue
Hydrozincite	white, light blue
Izokite	blue, orange
Kaolinite	blue, light blue, dull blue, bright deep blue, dark blue, none
Kyanite	red, dark violet-red
Magnesite	dull red
Nacrite	dull blue with bright specks
Pitchblende	none to orange
Rhodochrosite	dull orange, violet, bright red
Scheelite	intense CL, yellow, white, blue
Schroeckingerite	strong dark green
Sillimanite	dull blue, dull bluish-white, red
Smithsonite	bright blue
sphalerite	generally non CL, rarely light blue, dark blue, orange-red,
Spodumene	red, bright pale orange, bright orange, bright pink
Strontianite	blue, blue-violet, salmon pink, light orange
Tremolite	bright red
Uranopilite	bright green
Willemite	strong light green, green, violet
Wöhlerite	blue-green
Wollastonite	bright yellowish green, intense yellow
Wurtzite	green
Zircon	blue, yellowish, variety of other colours

2.10. INFRARED SPECTROSCOPY

Infrared adsorption spectroscopy is used to identify minerals that contain tightly bound molecular groups such as CO_2, SO_4, OH, etc. (e.g. anglesite ($PbSO_4$)). An infrared absorption spectrum is produced when a mineral is irradiated by radiation within the infrared range (wavelength ranging from about 1 μm to 1 mm) (Figure 2.4). The irradiation causes changes, which are specific for each mineral, in the vibrational energy of the constituent molecules in the minerals (Jones, 1987, Zussman, 1977). The changes are recorded as absorption bands at different wavelengths for each molecule group. Samples are prepared for analysis as pellets composed of about 1 mg of pulverized material and about 300 to 400 mg of KCl.

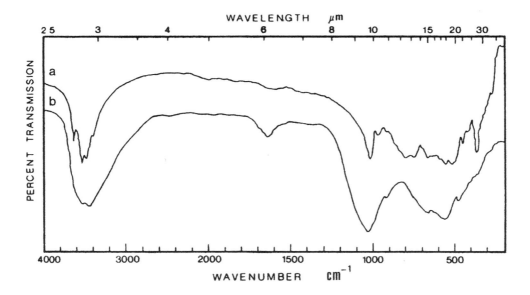

Figure 2.4. Example of an IR spectrum: (a) Gibbsite ($Al(OH)_3$) from Australia, (b) Gibbsite-like mineral ($Al(OH,F)_3$) from Francon Quarry, Montreal, Quebec. (From Jambor et al., 1990). The absorption bands in the high region of the spectrum relate to OH stretching.

2.11. IMAGE ANALYSIS SYSTEM

Image analysis is used in applied mineralogy to determine mineral quantities, liberations of minerals, size distributions of grains and particles, sizes and quantities of inclusions in particles, particle shapes, mineral associations, and other characteristics that relate to the morphology of particles. Image analysis techniques have been described in detail by Lastra et al. (1998b), but the highlights are summarized here because there is a paucity of publications on image analysis techniques. Samples of large pieces and/or of crushed and ground materials are mounted in polished, thin and/or polished-thin sections for image analysis, and samples of airborne dusts collected on polycarbonate membrane filters can be analysed directly. Since image analysis is generally performed on material that is heterogeneous, many fields of view (10 to several hundred) need to be analysed, preferably in an automatic unattended mode, to get statistically valid data.

Performing image analysis involves (1) image procurement, (2) image enhancement, (3) segmentation, (4) mineral identification, (5) preparation of image for measurement, (6) measurements, (7) data manipulation and (8) interpretation of results. The analyses are performed with an image analysis system that consists of (a) an instrument for image procurement (e.g. optical microscope, SEM, microprobe), (b) an image analyser to carry out the image analysis routines, and (c) a computer for data manipulation.

Image analysis can be performed satisfactorily when the image procurement instrument produces an image that displays enough differences between the minerals in the sample to enable segmentation into mono-mineralic images. The most commonly used image procurement instruments are optical microscopes, SEM and MP.

2.11.1. Optical microscope

Images produced with an optical microscope are transferred to an image analyser via a black and white and/or colour TV camera and a frame grabber. The current image analyser manufacturers subdivide the black and white images into 256 grey levels, with black designated as 0 grey level, and white as 255. If a mineral displays a unique grey level in the black and white image, or a distinct colour in the colour image, its image can be segmented from the image of the field of view. The image of the field of view is generally prepared for segmentation by enhancing the differences between the minerals with edge enhancement routines and equalization or averaging filters (e.g., sigma filter). The segmentation is performed by detecting and setting the lower and upper grey levels for each mineral to produce separate binary images. The binary image is an image whose grey level is 255.

The optical microscope is not a good imaging instrument because the grey levels of most minerals in reflected light are so close to each other that images of comparatively few minerals can be readily discriminated and segmented from the image of the field of view. Even fewer minerals can be discriminated in transmitted light images. Image analysis using colour microscopy is not significantly better, although a combination of colour and grey level images may provide some improvement. A second constraint with the optical microscope is that, due to the configuration of the objective lens, uniform lighting is difficult to achieve and maintain across the field of view. Hence a shading correction of the field background must be used. Unfortunately, the shading may change as the sample is moved to new fields of view because of different distributions of the minerals. Therefore, when the imaging instrument is an optical microscope, interactive editing is required, and automatic unattended analysis is not possible. Interactive editing involves modifying the binary image by completing the images of poorly detected grains, and erasing features that were incorrectly segmented due to poor image input of the field of view. It generally takes an operator about 15 to 20 minutes of interactive editing to correct an image of a field of view.

2.11.2. Scanning electron microscope (SEM) or microprobe (MP)

Images produced with a scanning electron microscope (SEM), microprobe (MP) and other microbeam techniques are transferred to an image analyser via an interface and a frame grabber as black and white images, generally referred to as grey level images.

2.11.2.1. Discrimination by grey levels (grey level images)
In contrast to the relatively few minerals that can be discriminated in images produced with an optical microscope, many minerals can be discriminated from each other in backscattered electron images (BSE) produced with a scanning electron microscope (SEM) or a modern electron microprobe (MP), and identified with an energy dispersive X-ray analyser (EDS). The BSE image is used because the grey levels of the features in the image are proportional to the average atomic number of the mineral (Table 2.1). Minerals with relatively small differences in average atomic number (0.5 to 1) can display grey levels that are different enough to be segmented from each other. The grey levels are generally uniform across the BSE image, hence shade correction is usually unnecessary. Furthermore, the grey levels for each mineral remain constant as the sample is moved from one field to another, provided that the beam current on the Faraday cup, and the GAIN and BRIGHTNESS on the BEI amplifier remain fixed (Petruk, 1989). Under these conditions, mineral identification via the grey level technique of the BSE image can be used with confidence. Automatic unattended image analysis is therefore possible, but the SEM needs a motorized stage that is controlled by the image analyser. The motorized stage has to be set (by software) to move precisely in X and Y directions, and calibrated so that the stage would move to adjacent, or uniformly spaced fields of view at different SEM magnifications.

To perform automatic image analysis using grey levels of minerals the lower and upper grey levels must be set manually for each mineral, and fixed for segmentation during automatic unattended analysis. The identities of the minerals corresponding to each grey level are determined interactively, prior to the automatic analysis, with the EDS.

Most image analysers can store images on diskettes, hard disk or a Bernoulli disk. A 512 x 512 image takes 262,144 bytes of image memory, i.e., nearly 4 images can be stored per megabyte of image memory. Collection of many images can be utilized when the IA uses a MP or SEM as the imaging instrument and all minerals can be discriminated and identified on the basis of grey level. If the image analyser has enough memory space to collect a large number of images, the images can be collected sequentially, and the MP or SEM would be free for other work while automatic image analysis is performed with the image analyser on the collected images.

The grey level image collected from the SEM or MP is prepared for segmentation into binary images by using edge enhancement and equalization routines to enhance the differences between the minerals displayed in the image.

Since the minerals are identified by the grey level, the minimum size of the particle that can be detected and identified, is dependent upon the magnification. Particles >1 μm in size can be detected, identified and segmented from images produced at magnifications of 400X.

2.11.2.2. Discrimination by mineral compositions (several minerals with same grey levels)
In some instances several minerals have the same average atomic number and grey level in the BSE image, but have different compositions. In these cases the minerals are discriminated from each other on the basis of composition rather than grey level. Discrimination between minerals of the same grey level is performed by (1) obtaining a multi-mineral binary image for the specific grey level, and (2) using a beam steering technique to raster the electron beam over each particle in the image to detect the X-ray signals produced by the different elements in the minerals. Two techniques can be used; one is a dot mapping technique, which produces dot maps for one or more unique elements that are specific to each mineral, and the other is an X-ray counts technique, which uses X-ray counts to determine the quantities of distinct elements in the minerals.

Two approaches are used for the **dot mapping technique**. One is to produce a dot map of the entire field (field dot-mapping technique), and is used when many particles are present in the multi-mineral binary image. The other is a window dot mapping technique which is used when only a few particles are present in the multi-mineral binary image.

The window dot mapping technique involves defining the outline of each particle or creating a box around each particle in the multi-mineral binary image, and producing dot maps within the defined areas. Less time is needed to map the smaller defined areas, than to map the entire field.

If there are many particles in the multi-mineral binary image, so much time is lost in moving from window to window that it is quicker to map the entire field.

The dot mapping technique produces different densities of dots. It is necessary to convert the areas with high densities of dots into a solid image that represents the mineral in each particle, and to eliminate the areas with low densities of dots, which are due to background noise. This is done with a rank or median filter of the image analyser. The resulting image is integrated with other images that were obtained from the sample by the grey level segmentation technique. The integration is performed by using a Boolean operator of the image analyser.

The **X-ray counts technique** can be used instead of the dot mapping technique to separate and identify the minerals in the multi-mineral binary image, and is the only technique that can differentiate between minerals that contain different proportions of the same elements (e.g. pyrite and pyrrhotite). However, each individual in the multi-mineral binary image must contain only one mineral, as individuals composed of two minerals within the same grey level (e.g. grains composed of chalcopyrite and pentlandite) cannot be identified by this technique. Nevertheless, they can be identified by a variant of the technique as discussed below. The technique involves:
- selecting the grey level that contains the minerals in the multi-mineral binary image,
- setting-up the windows in the EDS unit for the elements that are present in the minerals of the multi-mineral binary image (up to 8 elements for most EDS units),
- scanning with the electron beam, to acquire X-ray counts for each selected element in each grain displayed in the multi-mineral binary image,
- transferring to the image analyser, via a hardwire from each window, the X-ray counts for each element in each grain,
- comparing the X-ray data to a mineralogy table to identify the mineral in each grain,

- recording the results, and displaying them on the CRT screen by a different colour for each mineral.

A variant of the X-ray counts technique is to create equally spaced analysis points within each grain in the multi-mineral binary image, and to obtain X-ray data for each point. The distribution of minerals within each grain would be determined by this technique, and can be used to analyse grains that contain more than one mineral in each individual of the multi-mineral binary image. This routine has the same principle of analysis as the QEM*SEM.

2.11.3. QEM-SEM

The QEM*SEM was developed by CSIRO in Australia to analyse ores and mill products with respect to mineral processing. The instrument uses the BSE image to obtain particle outlines that serve as frames for analysis of each grain. The electron gun is steered within the frames to scan each particle and to obtain X-ray counts for 16 elements at designated pixel points within the particle. The system uses four EDS detectors. The X-ray counts for each element at each pixel position are sent to computers (several micro VAXES were used in 1989). The computers compare the X-ray counts to a reference mineral file to identify the mineral at each pixel point. Each identified point is recorded in a file and displayed on the CRT screen by a colour which represents the mineral. Calculations of the data with respect to mineral processing are performed automatically, and no further image analysis is performed (Pignolet-Brandom and Reid, 1988; Reid and Pignolet-Brandom, 1988) . It is noteworthy that since the mineral is identified by X-ray analysis at each designated pixel, the technique is relatively slow. Similarly since the generation volume for X-rays is around 3 to 5 μm, grains smaller than 5 μm cannot be identified with confidence, particularly for pixels near grain boundaries. On the other hand the QEM*SEM performs the analysis unattended with a high degree of confidence in the results.

2.11.4. Enhancement

The quality of grey level and binary images can be enhanced with filters to improve detection and discrimination between minerals. The filters can bring out mineral characteristics that will help to differentiate a mineral, and suppress characteristics that interfere. For example the use of an edge enhancement filter followed by a Laplace filter can bring out grain boundaries, and suppress the features within the grains.

2.11.5. Preparing binary images of minerals for measurement

It has been shown above that binary images of minerals are produced by:
- segmenting optical microscope images,
- segmenting grey level BSE images,
- using a dot-mapping technique,
- using an X-ray count technique.

These binary images generally contain small artifacts due to halo effects, to limits in segmentation, and to electronic noise. The artifacts must be removed before the images are analysed. The halos are narrow borders that appear around other minerals due to a transition from one grey level to the next. Other artifacts are small dots. Most image analysers have a routine that will remove all features that are smaller than about 3 pixels in diameter and in width, and restore the larger features to their original size and approximate shape. The routine is given a different name in each image analyser; it is called *Scrap* in the Kontron (Zeiss since 1998) image analyser. This routine might slightly modify the outlines of the grain boundaries. A *rank or median operation* performs similar image cleaning. It also modifies the outlines of the grain boundaries. Several other routines, which remove the artifacts and restore the original grain boundaries, have been written as macros by the author. One is *erode-dilate-(Boolean- and)*, and the other is *erode-mark object* (Petruk, 1989b). Details of the routines are described by Lastra et al. (1998b).

Some particles may be so close together that they touch. The image analyser sees touching particles as one particle, and would produce wrong results when performing object specific measurements such as size analysis and liberation analysis. It is possible to separate images of the touching particles if there is a narrow neck between the two particles. The operation is performed by eroding the particle images till they separate at the narrow neck and restoring them to their original size but keeping them separated at the neck. Alternately a border is drawn around all particles, and a grain boundary routine creates borders across touching particles.

2.11.6. Mineral quantities

Mineral quantities are determined using either an area or a linear intercept analysis. The area analysis involves counting the number of pixels that it takes to cover the area of the detected features in the binary images of each mineral. The linear intercept analysis involves measuring the intercept length across each feature in the binary images of each mineral.

2.11.7 Size distributions of minerals

Size distributions of particles and/or mineral grains are determined by measuring each feature using either an *area measurement* or a *linear intercept measurement* to define the proportion of particles or of mineral grains in each size range. The diameter of the feature is used to define the size. Several types of measurements can be used to define a grain or particle size; they include Dcircle, Dmean, Dsquare, and Dmax. The author generally uses Dsquare because it fits closest to sieve analysis.

Dcircle = calculated diameter by measuring the area of a feature and assuming that the feature is a circle.

Dmean = measuring several diameters (generally 16) on a feature and calculating the mean diameter.

Dsquare = measuring the area of a feature and calculating the diameter by assuming that the feature is a square.

Dmax = measuring several diameters (generally 16) on a feature and using the maximum.

The size analysis of mineral grains in a polycrystalline monomineralic mass can be determined by constructing grain boundaries within the image of the monomineralic mass. Grain boundaries can be constructed with image analysers that have the appropriate routines, and if the polished sections of the monomineralic mass display incipient features (fractures and pits) along grain boundaries. The incipient features are used as nuclei for constructing the grain boundaries in the image of the monomineralic mass. The grain boundary routine in the Kontron-IBAS image analyser is named *GRAINS*, and is described by Lastra et al. (1998b).

2.11.8. Grain shapes and grain boundary irregularities

Grain shapes are defined by measuring the features in binary images of each mineral to obtain values for such parameters as area/perimeter, aspect ratio, circularity, etc. These measurements are performed directly on binary images of each mineral.

Grain boundary irregularities can be measured only with an image analyser that has a *binary thinning* routine or its equivalent. The *binary thinning* routine erodes a feature till it is 1 pixel wide, then stops eroding. The routine for measuring grain boundary irregularities involves obtaining an image of the grain boundary, and measuring its length. The image of the grain boundary is then dilated several cycles, followed by binary thinning which produces a straight boundary. The straight boundary is measured and compared to the original boundary to define the grain boundary irregularity.

2.11.9. Liberations of minerals

Mineral liberations are determined by measuring the area of the host particle and of the inclusion in the host particle and calculating the area percent of inclusion/area percent of particle as degree of liberation. Mineral liberation is discussed in Chapter 1.

2.11.10. Mineral associations

Mineral associations and measurement techniques are discussed in Chapter 1.

2.11.11. Applications of image analysis to on-line control

Applications of image analysis to on-line control in mineral processing and mining have been investigated for some time (Petruk, 1983b; Morrison, 2000). Recent publications show that research has been done to:
- infer mineral characteristics that affect concentrator operations during mineral processing by obtaining images of the froth in flotation cells and using textural analysis to interpret the images (Bezuidenhout et al., 2000; Bonifazi et al., 1997a; Bonifazi et al., 2000),
- develop a technique for selective mining of oil sands and industrial minerals by determining zones of ore-grade material at the mining face using short-wave infrared to detect the distribution of wanted minerals and gangue (Sedgwick et al., 2000),
- determine the size distributions of blasted and crushed materials using a video camera at the Keewatin, Minnesota operation of the National Steel Pellet Co. (Herbst, 2000),
- develop an apparatus (Manybarycentre Classifier) for separating glass on the basis of colour, by using an optical system connected to a colour video camera (Bonifazi et al., 1997b). Bonifazi et al. (1997b) separated glass fragments into five colour classes (light green, dark green, white, half white and brown).

CHAPTER 3

MINERALOGICAL CHARACTERISTICS AND PROCESSING OF MASSIVE SULFIDE BASE METAL ORES FROM THE BATHURST-NEWCASTLE MINING AREA

3.1. INTRODUCTION

Many volcanogenic massive sulfide (VMS) base metal deposits occur in the Bathurst-Newcastle mining area in New Brunswick, Canada. Some deposits are presently being exploited, some are in the developmental stage, some have been tested and are currently dormant, some have been mined out, and others are in various stages of exploration. The ores in these deposits are fine-grained, and it is difficult to obtain high metal recoveries from most of the ores by mineral processing techniques. The metals are recovered in at least three concentrates, zinc, lead and copper. Silver and sometimes gold are recovered as by-products in the lead and copper concentrates. Mineralogical investigations of the ores have helped mineral processing engineers design and/or modify flowsheets for better recoveries of saleable grade products. The ore characteristics and mineralogical properties that affect mineral processing of the ores from the Bathurst-Newcastle area are discussed in this chapter. It is noteworthy that many characteristics of the deposits in the Bathurst-Newcastle area are similar to the characteristics of the VMS deposits in the Iberian pyrite belt in Portugal and Spain, and similar problems are experienced in processing the Portuguese and Spanish ores (Gaspar, 1991; Morales, 1986; Badham, 1982).

3.2. GENERAL CHARACTERISTICS OF BATHURST-NEWCASTLE BASE METAL DEPOSITS

The volcanogenic base metal deposits in the Bathurst-Newcastle mining area in New Brunswick, Canada, are composed of weakly to moderately metamorphosed massive sulfides, and display the following characteristics of sphalerite-galena VMS deposits:
- They consist of pyrite-rich massive sulfide bodies that contain variable amounts of sphalerite, galena and chalcopyrite.
- Some zonation is present, grading from a pyrite-chalcopyrite-pyrrhotite zone at the footwall, through a pyrite-sphalerite zone in the main parts of the deposits, to a pyrite-sphalerite-galena zone in the upper parts.
- The pyrite-chalcopyrite-pyrrhotite zone is narrow in most deposits although it was relatively wide in the Wedge deposit, which is now mined out.
- The pyrite-sphalerite zone is commonly banded and displays sphalerite-rich and pyrite-rich bands.
- A feeder zone, composed of stringers and disseminations of chalcopyrite and pyrrhotite in chloritic rocks is present at most deposits, and forms the major mineralized zone at the Chester deposit (Adair, 1992; Hamilton, 1992; Irrinki, 1992; Jambor, 1979; Luff et al., 1992; McCutcheon, 1992; Rennick and Burton, 1992; Whaley, 1992; Petruk, 1959).

3.2.1. Mineralogical characteristics

Sulfide minerals commonly make up 70 to 90 volume percent of the massive sulfide bodies; the remainder is mainly quartz, chlorite and sericite with minor to trace amounts of feldspar, biotite, carbonates (siderite, dolomite, and calcite), amphibole, pyroxene, monazite, barite, apatite, rutile, titanite, and other silicate minerals. Pyrite is the major sulfide mineral in the massive sulfide bodies. Sphalerite and galena are the main minerals of economic value, and chalcopyrite and tetrahedrite-freibergite are of secondary economic value. Chalcopyrite was, however, the main mineral of economic value in the copper-rich zone of the Wedge deposit. The volcanogenic massive sulfide ores in the Bathurst-Newcastle mining area contain, in wt %, about 2 to 15 Zn, 1 to 5 Pb, and 0.2 to 3 Cu; in ppm, about 50 to 150 Ag, 0.2 to 1.7 Au; and minor to trace amounts of Cd, Mn, Sn, Sb, As, Co, Bi, In, Se, and Te.

3.2.1.1. Pyrite

The pyrite (FeS_2) generally occurs as masses of relatively fine-grained pyrite. The massive pyrite contains sphalerite (ZnS), galena (PbS), chalcopyrite ($CuFeS_2$), quartz, carbonates, and tetrahedrite-freibergite ($(Cu,Ag)_{10}(Fe,Zn)_2(Sb,As)_4S_{13}$) as pseudoveinlets and irregular grains in interstices between pyrite grains and between clusters of pyrite grains (Figure 3.1), and as narrow veinlets in fractured pyrite. Arsenopyrite (FeAsS), magnetite (Fe_3O_4), and pyrrhotite ($Fe_{1-x}S$), locally occurr as major components in the massive pyrite. Some comparatively large grains of recrystallized pyrite are present, and a few of the recrystallized pyrite grains contain rounded inclusions of trapped minerals, mainly galena and sphalerite. In addition remnants of framboidal pyrite were observed in the Caribou deposit (Jambor, 1981; Chen, 1978).

Figure 3.1. Photomicrograph showing massive pyrite (white) with sphalerite (grey) in the interstices. The black grains are silicate minerals.

3.2.1.2. Sphalerite

Up to 75 % of the sphalerite in the pyrite-sphalerite zone occurs in massive sphalerite bands that range from about 1 cm to 15 cm in width, and grade into wider bands of massive pyrite. The massive sphalerite contains inclusions of euhedral pyrite, rounded to irregular galena grains, irregular tetrahedrite-freibergite grains, and small grains of cassiterite, magnetite, chalcopyrite, quartz, and carbonates (Figure3.2a, 3.2b). The remaining sphalerite in the pyrite-sphalerite zone, and most of the sphalerite in the pyrite-sphalerite-galena zone, occurs as interstitial fillings in massive pyrite (Figure 3.1). The sphalerite from the Brunswick No. 12, Brunswick No. 6 and Heath Steele deposits contains an average of about 59.5 wt % Zn, 7 wt % Fe, low Cd (~0.1 wt %) and low Mn (<0.1 wt %) (Petruk and Schnarr, 1981; Chen and Petruk, 1980; Lastra et al., 1995), and about 22 ppm Ag (Chryssoulis et al., 1985). Micro PIXE analyses of 40 sphalerite grains in the rougher feed to the Zn circuit of Brunswick Mining and Smelting gave, in ppm, an average of 142 Ga, 144 As, 260 In, 33 Sn and 1524 Pb (Lastra et al., 1995). The Fe content in the sphalerite varies from about 2 to 9 wt % Fe, and visual observations of the ores from the Brunswick No. 12 deposit show that about 10 % of the sphalerite is the "blonde" variety, the remainder is marmetitic (e.g. brown to dark brown). Microprobe analyses of 105 sphalerite grains from the Zn concentrate of Brunswick 12 ore show that 10 grains (9.5 %) contained 2 to 3 wt % Fe (blonde variety) the remainder contained 3 to 9 wt % Fe (Lastra et al., 1995).

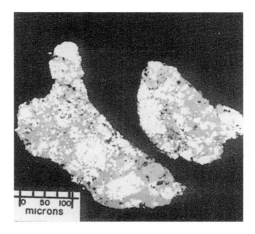

Figure 3.2a. Massive sphalerite band (dark grey) showing inclusions of pyrite (very light grey), galena (light grey), cassiterite (outlined), and gangue (very dark grey and black).

Figure 3.2b. Particles from the interface between massive sphalerite and massive pyrite bands. They display massive sphalerite (grey) with numerous inclusions of euhedral and massive pyrite (white).

3.2.1.3. Galena

The galena occurs as irregular grains in massive sphalerite (Figure 3.2a), as interstitial and fracture fillings in massive pyrite (Figure 3.3a), as inclusions in pyrite (Figure 3.3a) and rarely as massive galena. The galena grains that occur as interstitial fillings and as inclusions in pyrite are small (generally ~1 to 10 μm in size), and the galena grains in vuggy pyrite are very small (ranging from <1 μm to several μm in size) (Figure 3.3b). The galena contains about 1 to 20,000 ppm Ag, and the Ag-rich galena grains contain up to 2.5 wt % Bi (Petruk and Schnarr, 1981; Chen and Petruk, 1980; Laflamme and Cabri, 1986b). The average Ag content in galena from the Brunswick No. 12 deposit is 800 -840 ppm, from the Heath Steele deposit 1400 ppm, and from the Caribou deposit 1080 ppm (Table 1).

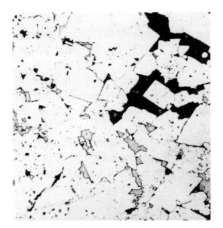

Figure 3.3a. Photomicrograph showing galena (grey) in interstices between pyrite (very light grey) and as minute inclusions in pyrite. The black areas represent silicate minerals.

Figure 3.3b. High magnification photomicrograph of two grains of vuggy pyrite (light grey) with minute (>1 μm) galena (grey) inclusions in vugs.

3.2.1.4. Chalcopyrite

The chalcopyrite occurs as interstitial fillings between pyrite grains (Figure 3.4), as small masses in massive pyrite, as veinlets (including hairlike veinlets) in fractured pyrite, as masses and discrete grains in the chalcopyrite-rich zones, and as disseminated chalcopyrite in the rock in the feeder zones.

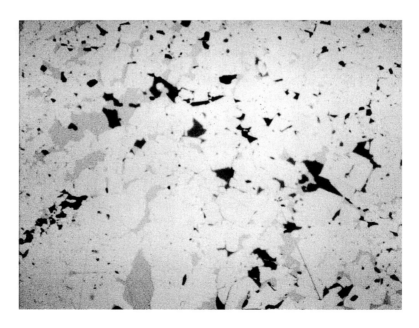

Figure 3.4. Chalcopyrite (grey) in interstices between pyrite grains in massive pyrite (light grey).

3.2.1.5. Silver

Most of the silver occurs as a minor to major element (trace to 35 wt %) in tetrahedrite-freibergite and as a trace element in galena. The remaining silver occurs as a constituent of pyrargyrite (Ag_3SbS_3), stephanite (Ag_5SbS_4), miargyrite ($AgSbS_2$), acanthite (Ag_2S), andorite ($PbAgSb_3S_6$), owyheeite ($Ag_2Pb_7(Sb,Bi)_8S_{20}$), jalpaite (Ag_3CuS_2), pyrostilpnite (Ag_3SbS_3), diaphorite ($Pb_2Ag_3Sb_3S_8$) and several unidentified sulphosalts, and as a trace element in chalcopyrite, sphalerite and pyrite. The average Ag contents in tetrahedrite-freibergite, galena, chalcopyrite, sphalerite and pyrite from the Brunswick No.12, Heath Steele and Caribou deposits are given in Table 3.1. It is noted that the average Ag contents in tetrahedrite-freibergite (top row, Table 3.1) reflect the Ag contents in the respective deposits (bottom row, Table 3.1) whereas the average Ag contents in galena do not.

Table 3.1
Average silver contents in minerals

Mineral	Brunswick No. 12	Heath Steele	Caribou
Tetrahedrite-freibergite (wt %)	16.4[1]	12.4[2]	7.2[3]
Galena (ppm)	840[4]	1400[2]	1080[3]
	800[7]		
Chalcopyrite (ppm)	590[5]		
Sphalerite (ppm	22[5]		
Pyrite (ppm)	7[5]		
Ag assay of ore (ppm) [6]	117	90	58

[1] = Owens, 1980; [2] = Chen and Petruk, 1980; [3] = Jambor and Laflamme, 1978; [4] = Petruk and Schnarr, 1981; [5] = Chryssoulis et al., 1985; [6] = Petruk and Wilson, 1993; [7] = Laflamme and Cabri, 1986a.

About 93 % of the Ag in the Brunswick No. 12 deposit occurs in tetrahedrite-freibergite + galena + chalcopyrite + sphalerite + pyrite (e.g. 52 % in tetrahedrite-freibergite, 35 % in galena, and 6 % in pyrite plus sphalerite plus chalcopyrite). The remainder is in silver sulphantimonides. In contrast, only about 75 % of the Ag in the Heath Steele deposit is in tetrahedrite-freibergite + galena + sphalerite + chalcopyrite + sphalerite + pyrite. The remaining 25 % of the Ag is in a large variety of Ag sulphantimonides (stephanite, pyrargyrite, cosalite, etc.) (Chen and Petruk, 1980). It is noted that the Heath Steele deposit is metamorphosed slightly more than the Brunswick No. 12 deposit.

The behaviour of silver during processing of the massive sulfide base metal ores from the Bathurst-Newcastle mining area is indicated in Table 3.2 by the distribution of Ag among the minerals in the products from the concentrator of Brunswick Division of Noranda (formerly Brunswick Mining and Smelting (BMS)). It is noted that 35 % of the silver in ore occurs as a trace element in galena. About 21 % was recovered (Dist.) as Ag in galena in the Pb concentrate which recovered about 65 % of the Pb in the ore.

Table 3.2.
Grades and distributions of silver in products from BMS concentrator, Sept., 1977

PRODUCT	Ag in galena		Ag in te, cp, pr, st, py, etc.		Total Ag	
	ppm	Dist.	ppm	Dist.	ppm	Dist.
Cu conc	149	**0.5**	4591	**16.3**	4740	**16.8**
Pb conc	339	**20.7**	291	**17.7**	630	**38.4**
Zn conc	23	3.3	57	8.1	80	11.4
Bulk conc	200	1.5	270	1.9	470	3.4
Sec. Zn conc	48	0.5	112	1.0	160	1.5
Tailings	11	8.3	27	20.2	38	28.5
Feed	42	34.8	75	65.2	117	100

te = tetrahedrite-freibergite, cp = chalcopyrite, pr = pyrargyrite, st = stephanite, py = pyrite.
Note: The distributions (Dist.) of recovered silver are shown in bold.

3.2.1.6. Gold

Most of the massive sulfide deposits in the Bathurst-Newcastle mining area have relatively low gold contents. The gold contents in the ores of Brunswick No. 12, Brunswick No. 6, and Heath Steele are generally <1 ppm, and nearly all of the gold is presumed to be present as invisible gold in pyrite and arsenopyrite. Most is lost to tailings with pyrite and arsenopyrite during mineral processing. Many searches for gold in polished sections have been conducted on samples from the Brunswick No. 12, Brunswick No. 6 and Heath Steele deposits, but only two gold grains were found in samples from the Brunswick No. 12 deposit and none from the Brunswick No. 6 and Heath Steele deposits. The gold grains in the Brunswick No. 12 deposit occurred as minute inclusions of electrum trapped in recrystallized pyrite in the chalcopyrite-rich zone (Leblanc, 1989) which forms a very small part of the deposit. In contrast, the Caribou deposit contains about 1.5 ppm Au, and electrum was found as minute grains in galena and chalcopyrite, and along pyrite grain boundaries (Jambor, 1981).

Most of the massive sulfide deposits in the Bathurst-Newcastle mining area were covered with a gossan that consisted of goethite and some silicate minerals, and was enriched in gold (Boyle, 1995). The gossan overlaid a narrow zone that was enriched in secondary Cu minerals (covellite, bornite, chalcocite, etc) and graded into unaltered massive sulfides. It is interpreted that the gossan was formed by oxidation which dissolved the sulfide minerals and released the constituent elements, including the invisible gold that was in the pyrite and arsenopyrite. Some of the dissolved Fe precipitated as Fe oxides (mainly goethite), whereas most of the dissolved S, Zn, and Pb were washed away. Some of the Cu precipitated as secondary Cu sulfides at the base of the gossan zone (Boyle, 1995), and most the Au precipitated in the gossan as secondary gold compounds. It is presumed that the gold that occurred as gold grains in the massive sulfides did not dissolve and remained as gold grains in the gossan. It is noteworthy that no one has reported finding gold grains in the gossan, but the author found an irregular blob, about 30 μm in diameter, of a gold compound in the Murray Brook gossan. The gossan was prepared for analysis by dissolving the silicates with HF and mounting the residue in a polished section. The polished section was studied with a SEM equipped with EDX. The X-ray signal from the blob, detected by EDX, displayed a broad Au peak which was about 1/2 as high as the peak for pure gold. Unfortunately it could not be determined whether the blob contained oxygen and/or carbon because a standard EDX detector, which did not detect elements with low atomic numbers, was used. The blob evaporated under the electron beam, which was operating at 20 kV and 15 nA. Some gossans, such as the Murray Brook gossan, were mined for gold and processed by direct cyanidation. A high gold recovery was obtained from the Murray Brook gossan by Nova Corp. (Boyle, 1995).

3.2.1.7. Trace minerals

Trace metallic minerals in the deposits are cassiterite (SnO_2), stannite (Cu_2FeSnS_4), pabstite $(Ba(Sn,Ti)Si_3O_9)$, bournonite ($PbCuSbS_3$), meneghenite ($Pb_{13}CuSb_7S_{24}$), boulangerite ($Pb_5Sb_4S_{11}$), cosalite ($Pb_2Bi_{12}S$), pyrargyrite (Ag_3SbS_3), stephanite (Ag_5SbS_4), miargyrite ($AgSbS_2$), acanthite (Ag_2S), andorite ($PbAgSb_3S_6$), owyheeite ($Ag_2Pb_7(Sb,Bi)_8S_{20}$), jalpaite (Ag_3CuS_2), pyrostilpnite (Ag_3SbS_3), diaphorite ($Pb_2Ag_3Sb_3S_8$), native antimony (Sb), native bismuth (Bi), bismuthinite (BiS), kobellite ($Pb_{22}Cu_4(Bi,Sb)_{30}S_{69}$), several unidentified sulphosalts, cobaltite (CoAsS), native Au (Au), electrum (Au:Ag = up to 20 wt % Ag) and marcasite (FeS_2) (Chen and Petruk, 1980; Petruk and Wilson, 1993; Jambor, 1981; Chryssoulis et al., 1985).

3.3. APPLIED MINERALOGY STUDIES OF BRUNSWICK No. 12 AND BRUNSWICK No. 6 ORES

The ores and mill products processed by the concentrator of Brunswick Division of Noranda (formerly Brunswick Mining and Smelting (BMS)) were studied intermittently from 1977 to 1994 to help the mineral processing engineers optimize metal recoveries. The first sampling campaign was conducted September 22 to 27, 1977 by sampling individual circuits over a two hour period when the concentrator was operating at steady-state conditions. At that time the BMS concentrator processed the ore in three parallel lines. Each line consisted of a Cu-Pb, a Zn and a secondary Zn circuit (Figure 3.5). The Cu-Pb circuit was composed of rougher, scavenger and cleaner circuits. Similarly the Zn circuit was composed of rougher, scavenger and cleaner circuits. The Cu-Pb concentrates from the Cu-Pb circuits of the three parallel lines were combined and processed in Cu separation and Pb dezincing circuits to produce Cu, Pb and bulk concentrates (Figure 3.5). The Zn concentrates from the Zn circuits of the three parallel lines were combined and upgraded in a Zn upgrading circuit by heating the Zn concentrate and using reverse flotation to remove the pyrite. The tails from the Zn circuits were re-processed in a secondary Zn flotation circuit to recover more sphalerite. Reverse flotation was also used to upgrade the secondary Zn concentrate. Samples were taken from each point in the circuits in line 3, and from the Cu separation, Pb dezincing, Zn upgrading and secondary Zn flotation circuits. Eighty-two samples were collected and studied at CANMET during 1977 to 1980. At that time line 3 processed about 3500 TPD of ore from the Brunswick No. 12 and Brunswick No. 6 deposits. Lines 1 and 2 were each processing about 3175 TPD of ore from the Brunswick No. 12 deposit. The grade of the feed to line 3 during the sampling campaign is given in Table 3.3.

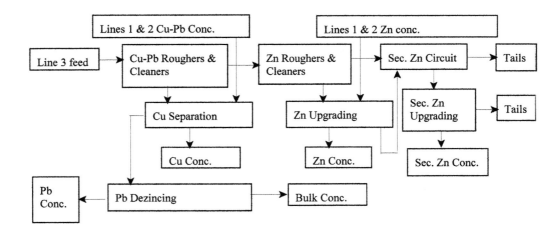

Figure 3.5. Simplified flotation circuit of BMS concentrator in 1979.

Table 3.3
Assay of feed* to line 3 of concentrator, Sept. 1977

Cu wt %	Pb wt %	Zn wt %	Fe wt %	As wt %	Ag ppm	Sb ppm	In ppm	Bi ppm	Sn ppm	Hg ppm
0.18	4.49	9.03	28.71	0.19	105	500	70	60	980	9

* from Petruk and Schnarr, 1981.

The ore was processed by grinding the concentrator feed to about 80 % minus 75 μm and regrinding some streams in open circuit, primarily to "repolish" the surfaces of the particles. The grades of the concentrates and tails and recoveries of the metals during the sampling campaign are given in Table 3.4. It is noteworthy that the grades and recoveries for both Zn and Pb were higher than was normally obtained by the concentrator in 1977. Hence the grades and recoveries for a one year period (August 8, 1978 to August 8, 1979), when the concentrator was operating in the same manner as in 1977, are included in table 3.4 and the values are considered as normal for that period.

3.3.1. Size distributions and mineral liberations

The size distributions of the sphalerite, galena and pyrite in unground ore, obtained in September, 1977, were determined by image analysis (Figure 3.6). The results showed that about 80 % of the sphalerite was in grains that were smaller than 78 μm, and about 80 % of the galena was in grains smaller than 32 μm. These size distribution data indicate that the ore must be ground to at least 80 % minus 78 μm to liberate about 50 % of the sphalerite (minimum grind for liberating sphalerite) and at least 80 % minus 32 μm to liberate about 50 % of the galena (minimum grind for liberating galena), using Petruk's liberation model (Petruk, 1986).

Table 3.4
Grades and recoveries in BMS concentrator*, Sept. 1977 and Aug. 1978 to Aug. 1979

Sample	Period covered	GRADE				RECOVERY/DISTRIBUTION			
		Pb wt %	Cu wt %	Zn wt %	Ag ppm	Pb %	Cu %	Zn %	Ag %
Pb Conc	Sept. 1977	**34.2**	0.7	8.9	**630**	**64.9**	14.2	7.3	**38.4**
Cu Conc	Sept. 1977	15.5	**22.0**	2.3	**4740**	1.5	**30.8**	0.1	**16.8**
Zn Conc	Sept. 1977	2.2	0.2	**53.6**	78	6.9	12.1	**77.8**	11.4
Bulk Conc	Sept. 1977	20.7	0.8	37.0	470	4.5	1.8	3.6	3.4
Sec Zn Conc	Sept. 1977	5.0	0.7	40.5	160	1.1	2.2	3.1	1.5
Tailings	Sept. 1977	1.2	0.15	1.3	40	21.1	38.9	8.1	28.5
Pb Conc	1978-1979	**31.5**	0.6	7.4	**475**	**60.1**	13.6	5.6	**34.4**
Cu Conc	1978-1979	7.3	**21.2**	3.1	**2892**	1.3	**41.9**	0.2	**18.9**
Zn Conc + Sec Zn Conc	1978-1979	2.0	0.2	**52.2**	81	7.3	7.9	**74.2**	10.9
Bulk Conc	1978-1979	18.6	0.7	33.7	353	7.2	3.0	5.2	5.2
Tailings	1978-1979	1.1	0.14	1.7	38	24.1	33.6	14.8	30.6

*from Petruk and Schnarr (1981).

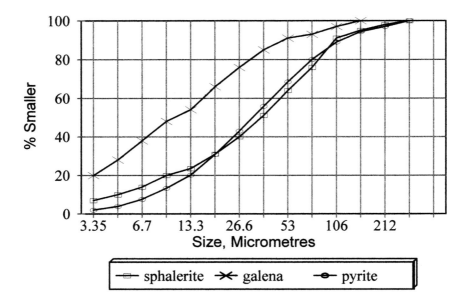

Figure 3.6. Size distributions of galena, sphalerite and pyrite in unground ore from the BMS concentrator feed (September, 1977), plotted as percent smaller than.

Liberation studies were performed by image analysis of mill products to determine the behaviour of the minerals during processing and to determine the optimum grind for liberating the minerals (Petruk and Schnarr, 1981). It was found that high recoveries were obtained for liberated sphalerite and for particles containing more than 70% sphalerite (e.g. >70% sphalerite particles), but not for unliberated sphalerite in <70 % particles. At the grind of 80% minus 75 μm the ore contained a significant amount of unliberated sphalerite in particles with less than 70% sphalerite which were large enough to liberate the sphalerite by regrinding (Petruk and Schnarr, 1981). Hence finer grinding was required to obtain optimum liberation, and a regrind of 80% minus 43 μm was recommended, even though the size distribution data indicated a finer grind. Grinding tests, conducted by Lakefield Research in the early 1980's, confirmed that regrinding to 80% minus 43 μm would increase Zn recoveries by several percent. Closed circuit regrind mills were installed in line 3 of the concentrator and improved recoveries were obtained. Since then many other modifications have been made to the concentrator to improve metal recoveries.

A series of grinding tests was conducted in 1987 (Petruk, 1988) by grinding samples of minus 10 mesh rod mill discharge for 15, 30, 45, 60 and 90 minutes in a ball mill using 2 kg samples. At these grinds the ore was reduced to 80 % minus 70 μm, 80% minus 40 μm, 80% minus 32 μm, 80 % minus 25 μm, and 80% minus 22 μm respectively (Figure 3.7). Mineral liberations in sieved fractions (150 to 106 μm, 106 to 75 μm, 75 to 53 μm and 53 to 37.5 μm) were measured by image analysis (Petruk, 1988). The sizes of the particles and the liberations of minerals in the particles in the unsieved minus 37.5 μm fraction were also measured by image analysis, by treating the minus 37.5 μm fraction as an unsieved sample (see Chapter 1).

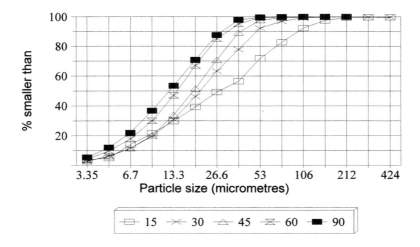

Figure 3.7. Size distributions of particles obtained by grinding samples of the ore for 15, 30, 45, 60 and 90 minutes. Plotted as percent smaller than.

The apparent liberations obtained for the 15, 30, 45, 60 and 90 minute grinds were 33 %, 34 %, 48 %, 52 % and 60 % respectively for liberated sphalerite, and 13 %, 13 %, 14 %, 15 % and 19 % respectively for liberated galena (Figure 3.8). Liberations for sphalerite in particles containing >70 % sphalerite were 80 %, 85 %, 88 %, 90 % and 90 % respectively, and for galena in particles containing >50 % galena were 71 %, 75 %, 75 %, 75 % and 86 % respectively (Figure3.8) (Petruk, 1988a).

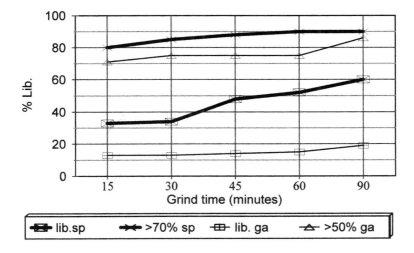

Figure 3.8. Liberations of sphalerite (sp) and galena (ga) in ground products.

The liberation data for the >70 % sphalerite particles and for the >50 % galena particles, in material from all grinds, were plotted as cumulative percent in sized particles, beginning with the largest size and continuing to the finest size (for each grind grind) (Figure 3.9). The plots for particles containing >70 % sphalerite show that sphalerite liberations (as >70 % particles) begin between 300 and 75 μm (15 minute grind), begin to increase for particles between 75 and 37.5 μm (all grinds), continue to increase to 9.4 μm particles, and reach a maximum at 4.7 μm particles (maximum liberation is 90 % for >70 % particles). It is interpreted from these data that the minimum grind for liberating sphalerite is about 80 % minus 75 μm (as determined in the initial study in 1977-80), and the optimum liberation for liberating sphalerite (in particles containing >70 % sphalerite) was reached in the 45 minute grind sample (80 % minus 32 μm). The optimum grind of 80 % minus 32 μm is selected because the liberations of >70 % particles did not increase significantly for samples that were ground for longer periods of time. The interpretation was partly confirmed by pilot plant and concentrator tests conducted in 1991 (Shannon et al., 1993). The tests showed major improvements in Zn recoveries when the ore was ground to around 80 % minus 30 μm. Subsequent tests on the cleaner circuit of the concentrator showed that the maximum Zn recovery was obtained from the cleaner cells when the rougher concentrate was reground to about 80 % minus 20 μm. However, a predominance of material between 8 and 30 μm was required in the flotation feed (Shannon et al, 1993) and it was necessary to use a low density slurry (~15 %). Zinc losses increased when the density of the slurry increased and/or when the quantity of particles smaller than 8 μm increased. Unfortunately, it was not practical at the time to use such a low density slurry in the concentrator (Shannon et al., 1993), and such a fine grind would undoubtedly produce large quantities of minus 8 μm material.

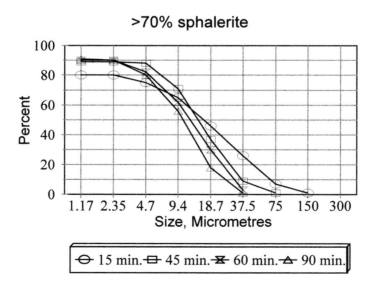

Figure 3.9. Cumulative percent of >70 % sphalerite particles with respect to size ranges.

The plots for particles containing >50 % galena (Figure 3.10) show that galena liberations (in >50 % particles) begin between 150 and 75 μm (15 minute grind and RMD), begin to increase for particles between 37.5 and 18.7 µm (all grinds), continue to increase to 4.7 µm particles for the 15, 45 and 60 minute grinds, and continue to increase to 1.17 µm particles for the 90 minute grind (maximum liberation was 86 % for >50 % particles). It is interpreted from these data that the minimum grind for liberating galena is about 80 % minus 37.5 µm (about the same as determined in the initial study in 1977-80), but the maximum (optimum) grind for galena liberation was not reached. This interpretation was confirmed by a research study conducted by Grant et al. (1991) on optimizing grind for galena recovery in the concentrator. They found that the highest Pb recoveries were obtained in sized fractions between 30 and 8 µm. Subsequent studies have shown that Pb losses below 9 µm are due to slime losses of galena (see below).

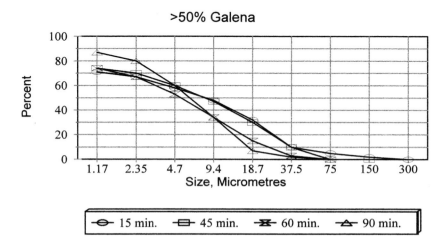

Figure 3.10. Cumulative percent of >50 % galena particles with respect to size ranges.

3.3.2. Texture Analysis

Since some of the sphalerite, galena and chalcopyrite occurs in interstices between pyrite grains and clusters of pyrite grains, and since some galena occurs as minute inclusions in pyrite, it is suggested that liberating these minerals involves breaking the pyrite into individual particles and releasing the interstitial material and inclusions (Smith and Petruk, 1988). It was observed that when the ground material was sieved the pyrite content was reduced in the coarser-grained sieved fractions. The reduction of pyrite contents occurred in successively smaller sized fractions as the size of the ground material decreased (e.g. ground for longer periods of time) (Figure 3.11). This observation suggests that the sizes at which the pyrite content is reduced represent breakage sizes for the pyrite. The reduction of pyrite content in sized fractions occurs because breakage of large pyrite masses produces smaller sized pyrite particles which are removed from the coarse-grained fractions by sieving. Figure 3.11 shows that the pyrite content decreased in the following sieved fractions: 104 to 150 (mid-point = 124) µm, 104 to 150 (mid-point = 124) µm, 73.7 to 104 (mid-point = 88) µm, 36.8 to 52.1 (mid-point = 44) µm and 36.8 to 52.1 (mid-point = 44) µm for the 15, 30, 45, 60 an 90 minute grinds, respectively. This indicates that the pyrite broke apart into particles that are 104 to 150 µm, 73.7 to 104 µm and 36.8 to 52.1 µm. Statistical studies by image analysis of the sizes, shapes, and surface roughness of the pyrite grains showed that the pyrite occurs as grains about 10 and 40 μm in size and as clusters of grains around 75 to 150 μm. These grain and cluster sizes correspond to the breakage sizes noted above for the pyrite, and to the liberation sizes for sphalerite in particles containing >70 % sphalerite (e.g. liberation begins between 300 and 75 μm and begins to increase at 75 to 37.5 μm). Since the galena occurs in smaller interstitial spaces than the sphalerite, galena liberation occurs when the smaller sized pyrite grains break apart.

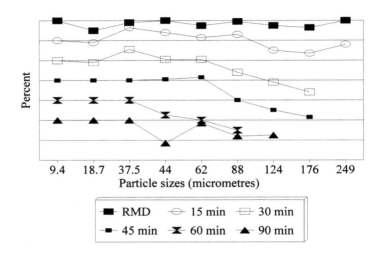

Figure 3.11. Relative pyrite contents in sized fractions of material from 15, 30, 45, 60 and 90 minute grinds and in rod mill discharge (RMD). Particle sizes are plotted at the mid-point between each Tyler mesh sieve size. The values for the pyrite contents (Y axis) are not shown because each curve begins at 58 % pyrite at 9.4 μm. The interval between each horizontal line in the graph is 20 %. The pyrite contents were reduced from 58 % in finest grained fractions to 45, 26 and 21 % in the 176 μm fractions of the 15, 30 and 45 minute grind samples, respectively. Similarly the pyrite contents in the 60 and 90 minute grind samples were reduced from 58 % in the finest grained fractions to 32 % and 42 %, respectively, in the 88 μm fractions.

3.4. APPLIED MINERALOGY STUDIES OF PRODUCTS FROM THE Cu-Pb ROUGHER CIRCUIT IN BMS CONCENTRATOR

A team of researchers was assembled by the Noranda Technology Centre in 1994 to study the Cu-Pb rougher circuit of the Brunswick Mining and Smelting concentrator in order to determine the characterisitics that affect flotation of galena, sphalerite and pyrite. The ore mined at the time was from a considerably lower part of the Brunswick No. 12 orebody, than the ore that was mined in 1977. Hence ore for the 1994 study will be referred to as from the lower part of the orebody, and for the 1977-80 study as from the upper part. The grade of the ore was essentially the same in the lower part as in the upper part; viz, the Cu-Pb rougher feed contained 9.5 wt % Zn, 3.5 wt % Pb, 0.22 wt % Cu and 113 ppm Ag. In contrast, the grain size was somewhat larger than in the upper part of the orebody. At a grind of about 80 % minus 43 μm (minus 325 mesh) the liberation was 56 % for apparently liberated sphalerite and 90 % for sphalerite in >70 %

particles. The liberations at the equivalent grind for the ore from the upper part of the orebody was 34 % and 84 % respectively. The liberations for galena at a grind of 80 % minus 43 µm were 46 % for apparently liberated galena and 68 % for galena in >50 % particles. The liberations at the equivalent grind for the ore from the upper part of the orebody were 13 % and 75 % respectively. The larger liberation sizes for sphalerite suggest that it should be possible to obtain somewhat higher Zn recoveries than was obtained from the ore in the upper part of the orebody. Similarly the larger liberation sizes for galena suggest that it should be possible to obtain higher grade Pb concentrates.

It was observed that too much of the sphalerite and pyrite in the ore floated in the Cu-Pb rougher cells, and some subsequently reported in the Cu, Pb and bulk concentrates. Similar observations had been made for the ore from the upper parts of the orebody (Table 3.4). The characteristics that affect flotation of sphalerite, pyrite and galena were studied.

3.4.1. Sphalerite

The Cu-Pb rougher feed contained 15.5 wt % sphalerite, and 15.8 % of it (e.g. 2.4 wt % sphalerite in the Cu-Pb rougher feed) was recovered in the Cu-Pb rougher concentrate which contained 16.4 wt % sphalerite. Image analysis studies showed that 32 % of the sphalerite in the Cu-Pb rougher concentrate was apparently liberated, and most of the unliberated sphalerite grains in the concentrate contained minute attachments of galena. The particles in the Cu-Pb rougher concentrate that were composed of 80 to 99.9 wt % sphalerite contained less than 10 wt % galena, and particles with less than 80 wt % sphalerite contained more than 10 wt % galena. Most of the liberated sphalerite in the Cu-Pb rougher concentrate was in particles that were smaller than 26.5 µm, whereas the unliberated sphalerite was in particles that ranged from 6.6 to 75 µm (Petruk and Lastra, 1995).

Since the unliberated sphalerite was intergrown with galena, there was a higher tendency for unliberated sphalerite than for liberated sphalerite to be float in the Cu-Pb rougher cells. Consequently, although about 56 % of the sphalerite in the feed sample was apparently liberated, only 32 % of the sphalerite in the Cu-Pb rougher concentrate was apparently liberated. On the other hand, flotation of the apparently liberated sphalerite in the Cu-Pb rougher cells and subsequent recovery in the Cu-Pb rougher concentrate is undesirable.

To determine why the apparently liberated sphalerite was in the Cu-Pb rougher concentrate, the coatings on sphalerite grains in various mill products were analysed by LIMS and ToF-SIMS at the Advanced Mineral Technology Laboratory (AMTEL), London, Ontario, Canada, as part of the study conducted by Noranda Technology centre (NTC), (Chryssoulis and Kim, 1994; Chryssoulis et al., 1995; Kim et al., 1995). It was determined that the apparently liberated sphalerite particles that floated in the Cu-Pb cells were coated with more Pb, marginally more Cu, less Fe, less Ca and less sulphates (e.g. oxidized less) than sphalerite particles that did not float. Other investigations, conducted at McGill university as part of the same study (Sui et al., 1996), as well as those at AMTEL, showed that Pb and Cu coatings activated the sphalerite particles, whereas the Fe, Ca and sulphate coatings did not. Furthermore it was determined that the activation by Cu coatings is about 10X as strong as activation by Pb coatings. It is considered noteworthy that a study of materials from the Zn circuit showed that dark (Fe-rich) sphalerite did not tend to float as readily in the Zn rougher cells, as blonde (Fe-poor) sphalerite (Chryssoulis et al, 1995). It is interpreted that this occurs partly because there is a higher amount of oxidized Fe on the surfaces of the dark sphalerite which reduces the Cu activation.

3.4.2 Pyrite

The feed to the Cu-Pb rougher circuit contained 50.2 wt % pyrite, and 16.7 % of it reported in the Cu-Pb rougher concentrate which contained 54 wt % pyrite. About 55% of the pyrite in Cu-Pb rougher concentrate was apparently liberated. Most of the unliberated pyrite that floated in the Cu-Pb rougher cells was in particles that contained galena. In particular, 34% of the pyrite in the Cu-Pb rougher concentrate was in particles that contained less than 10 wt% galena, and 11% was in particles that contained more than 10 wt% galena. The pyrite that floated was in particles of all sizes (Petruk and Lastra, 1995).

There was a higher tendency for particles containing unliberated pyrite with galena attachments than for liberated pyrite to float in the Cu-Pb flotation cells. However, since 63% of the pyrite in the feed was apparently liberated, most of the pyrite that reported in the Cu-Pb rougher concentrate (55% of the pyrite in the Cu-Pb rougher concentrate) was apparently liberated. ToF-SIMS analyses have shown that the apparently liberated pyrite which floated was coated with marginally more Pb, detectably more Cu, more Zn, and less sulphates (e.g. oxidized less) than the pyrite which did not float (Chryssoulis and Kim, 1994; Chryssoulis et al., 1995; Kim et al., 1995).

3.4.3. Galena

About 46 % of the galena in the Cu-Pb rougher feed was apparently liberated, and 68 % was in particles that contained >50 wt % galena. The galena liberation in the Cu-Pb rougher concentrate was 61 % apparently liberated, and 84% in particles that contained >50 wt % galena. Most of the apparently liberated galena particles in the Cu-Pb rougher concentrate were smaller than 26.5 μm (Petruk and Lastra, 1995). The liberation of galena in the Cu-Pb rougher concentrate increased as the particle size decreased (Figure 3.12), and the average galena content in particles of different size ranges increased as the particle sizes decreased (Figure 3.13). It is noted that the average galena content in the Cu-Pb concentrate (Figure 3.13) was very low, ranging from 11% in the 53 - 75 μm fraction to 32% in particles smaller than 3.4 μm. This is because the concentrate contained large quantities of sphalerite, pyrite, gangue and chalcopyrite in particles that did not contain galena.

Galena recoveries in the rougher concentrate were above 90 % for apparently liberated galena particles that were smaller than 37.5 μm. Recoveries of total galena increased with decreasing particle size from 71 % for the galena in 37.5 - 53 μm fractions to 92 % of galena in particles 4.7 to 13.4 μm (Figure 3.14) and dropped to 78 % for minus 3.4 μm particles (e.g. slimes loss).

About 85 % of the galena lost to the Cu-Pb rougher tails was unliberated, and occurred in particles that ranged from 75 to 9.4 μm in size. The apparently liberated galena grains that did not float (15% of galena in tails) were smaller than 9.4 μm in size (e.g. slimes loss).

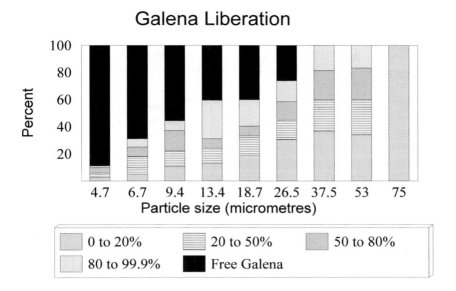

Figure 3.12. Proportions of galena with different degrees of liberation in particles (e.g. free galena, 80 to 99.9 % galena, etc.) of different size ranges in Cu-Pb rougher concentrate. The plotted particle size is the top size of sieved fractions and/or of size range determined by IA.

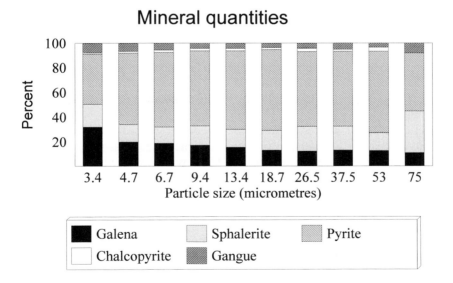

Figure 3.13. Average mineral quantities in particles of different size ranges in Cu-Pb rougher concentrate. The plotted particle size is the top size of sieved fractions and/or of size range determined by IA.

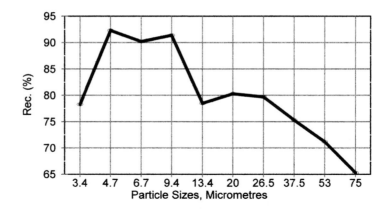

Figure 3.14. Recoveries of galena (free and unliberated) in Cu-Pb rougher concentrate in particles of different size ranges.

Since most of the galena-bearing particles in the concentrate ranged from 37.5 to 4.7 μm in size, it is interpreted that the minimum grind for galena liberation is about 80 % minus 37.5 μm. On the other hand, the increasing galena liberation with decreasing particle size indicates that extremely fine grinding (probably 80% minus 10 μm) would be required to liberate most of the galena. Unfortunately such a fine grind would produce a large amount of very fine-grained galena which would be lost as slimes. It is noted from the above paragraphs that a slime loss of apparently liberated galena begins for particles that are about 9.4 μm and becomes significant for minus 3.4 μm galena particles.

The unliberated galena in the concentrate tended to be associated with sphalerite when it was in particles that were larger than 13.4 μm, and with pyrite when it was in particles that were smaller the 13.4 μm (Petruk and Lastra, 1995). Figure 3.15 shows that the sphalerite content is higher in particles which are larger than 13.4 μm, and the pyrite content is proportionally higher in smaller particles. This indicates that much of the coarser grained unliberated galena (>13.4 μm) represents galena that was intergrown with sphalerite in the massive sphalerite bands. In contrast the fine-grained unliberated galena (<13.4 μm) represents the fine-grained galena that occurred as minute inclusions in both pyrite and sphalerite. This suggests that unliberated galena in particles that are larger than 13.4 μm might have different flotation characteristics than unliberated galena in particles smaller than 13.4 μm.

A grade-recovery curve, assuming perfect separation and recovery of galena-bearing particles of all sizes, was calculated for the galena in the Cu-Pb rougher feed by using the measured liberation data. The calculation was performed by using the quantities of galena and of galena-bearing particles in each particle category (e.g. particle grade) in the Cu-Pb rougher feed. The amounts recovered were assumed to be equal to the proportions of galena in the different particle grades (e.g. cumulative liberation yield), and the corresponding grade of the feed (e.g. wt% Pb) was calculated as the cumulative amount of galena at each point divided by the cumulative amount of galena bearing material. For example, since 46 % of the galena is apparently liberated, a recovery of 46 % would be obtained at a grade of 86.7 % Pb (e.g. Pb content in galena). The galena recovery for the second point (e.g. particles containing more than 90 wt % galena) is equal to the second point on the cumulative liberation yield curve, and the grade is ((46 + (quantity of galena in 90 - 99.9 % particles))/(46 + (quantity of 90 - 99.9 % particles)) * 86.7), etc. Figure 3.16 shows the grade-recovery curve for perfect separation and recovery of galena in Cu-Pb rougher feed. The graph also shows the grade and recovery of the Pb that was obtained in the Cu-Pb rougher concentrate. It is noted that the actual grade and recovery is far below the perfect separation curve because perfect separation was not obtained and the Cu-Pb contains large quantities of other minerals without attachments of galena.

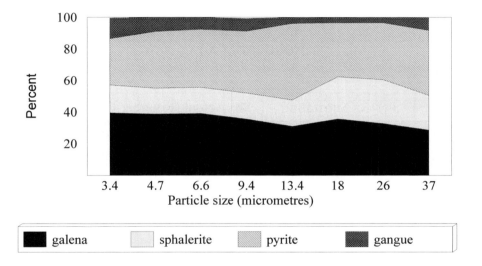

Figure 3.15. Mineral distributions in different sized particles containing 20 to 70 % galena in a Cu-Pb concentrate.

It was determined by studies at AMTEL and McGill university for NTC (Chryssoulis and Kim, 1994; Chryssoulis et al., 1995; Kim et al., 1995; Sui et al., 1996) that surface oxidation on galena begins when the ore is washed to keep the dust down, and continues during grinding and metallurgical processing. During oxidation, lead and sulfur are released from galena. Some of the released Pb migrates to the surfaces of other minerals, particularly sphalerite and pyrite, and causes their activation by collectors and results in undesirable flotation. It was also determined that the galena must be partly oxidized for xanthate attachment. Some of the lead and sulfur ions pass through several phases till they are converted to $PbSO_4$ (anglesite). Over oxidation (layer too thick) or conditions which lead to formation of $PbSO_3$ prevent xanthate attachment (Sui et al, 1996). Oxidizing conditions also lead to the formation of hydrophilic (non-floating) surface coatings such as oxides and hydroxides on these same minerals. As the oxidation process continues, sulfur begins to combine with the Pb and converts the surface coatings to sulphates which reduce the activation of the pyrite and sphalerite. Hence some of the pyrite and sphalerite that floated in the rougher flotation cells may not float in the recleaner cells.

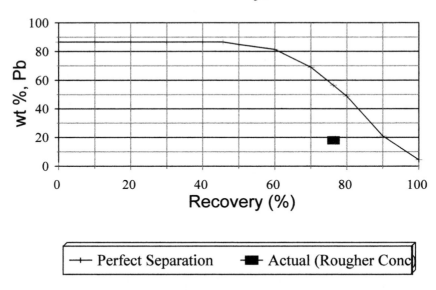

Figure 3.16. Grade-recovery curve for perfect separation and recovery of galena from Cu-Pb rougher feed.

3.5. SUMMARY

The following characteristics were determined for the volcanogenic massive sulfide deposits in the Bathurst-Newcastle mining area:

1. The deposits contain, in wt %, about 2 to 15 Zn, 1 to 5 Pb, and 0.2 to 3 Cu; in ppm, about 50 to 150 Ag, 0.2 to 1.7 Au; and minor to trace amounts of Cd, Mn, Sn, Sb, As, Co, Bi, In, Se, and Te.

2. The main ore minerals of economic value are sphalerite, galena and chalcopyrite.

3. Most of the silver occurs as a constituent of tetrahedrite-freibergite and galena and is recovered in Cu and Pb concentrates. Small amounts of silver occur in at least 10 Ag sulfides and sulphantimonides, and as a trace element in sphalerite, pyrite and chalcopyrite.

4. The main mineral characteristics that affect mineral processing are mineral liberations and coatings of Pb, Cu, Fe, and Ca on particle surfaces.

5. Grind sizes for liberating the sphalerite and galena from the ores of the Brunswick No.12 and Brunswick No. 6 deposits are: minimum and optimum grinds for liberating sphalerite are about 80 % minus 75 μm and 80 % minus 32 μm, respectively; minimum grind for liberating galena

is about 80 % minus 37.5 µm, the optimum grind is as fine as possible without producing too many slimes.

6. A method of producing grade-recovery curves, assuming perfect separation and recovery was developed.

7. The textures of pyrite influence ore breakage, and ore minerals that occur in interstitial spaces between pyrite grains tend to be liberated when the pyrite grains have been broken apart. Most of the pyrite occurs as clusters of grains that are 75 to 150 µm in size, and the clusters are composed of pyrite grains 10 to 40 µm in size. The ore, therefore, tends to break into particles around 80 % minus 75 µm, and on regrinding to particles about 80 % minus 32 µm.

8. About 15 % of the sphalerite and 15 % of the pyrite floats in the Cu-Pb rougher cells and is recovered in the Cu-Pb rougher concentrate. These minerals float in the Cu-Pb rougher cells because some of the unliberated sphalerite and pyrite is in particles that contain galena and chalcopyrite attachments, and some of the apparently liberated sphalerite and pyrite particles are coated with more Pb and Cu, and less Fe and Ca than those that do not float. The Pb and Cu are activators that cause the particles to float in the Cu-Pb roughers and the Fe and Ca suppress flotation of the particles.

CHAPTER 4

VOLCANOGENIC BASE METAL DEPOSITS IN THE FLIN FLON-SNOW LAKE AREAS, MANITOBA, CANADA

4.1. INTRODUCTION

The volcanogenic base metal deposits in the Flin Flon-Snow Lake areas in Manitoba, Canada occur in Proterozoic greenstone rocks which have been metamorphosed to the greenschist facies. The minerals in these deposits, therefore, display features of metamorphosed base metal ores. The characteristics of the Trout Lake and Callinan deposits in the Flin Flon area, and of mill products from the Trout Lake concentrator in Flin Flon, and the Chisel Lake and Stall Lake concentrators in the Snow Lake area, were studied by R. Healy through a Canada-Manitoba Mineral Development Agreement between the Canadian Federal Government, the Manitoba Government and the operating mining company (Hudson Bay Mining and Smelting), during the period of 1985 to 1990 (Healy and Petruk, 1988,1989a, 1989b, 1990a, 1990b, 1990c). W. Petruk was the scientific authority for the project. The Trout Lake and Stall Lake deposits are chalcopyrite-sphalerite deposits, the Callinan is a chalcopyrite-sphalerite-pyrite-pyrrhotite deposit, and the Chisel Lake is a sphalerite-galena-pyrite deposit. The characteristics of the Trout Lake and Callinan ores are summarized from the reports by Healy and Petruk (1988, 1990c) to highlight features of metamorphosed volcanogenic base metal ores. Similarly the characteristics of the mill products in the Trout Lake concentrator (Healy and Petruk, 1989a) are summarized to highlight the behaviour of a metamorphosed volcanogenic base metal ore during mineral processing.

4.2. CHARACTERISTICS OF TROUT LAKE AND CALLINAN ORES

The Trout Lake deposit consist of a series of lenses. Each lense exhibits the typical zonation of volcanogenic massive sulfides (Sangster, 1972; Franklin et al., 1981), and each is interpreted to represent an entire hydrothermal cycle. The ore types in the lenses progress from the hanging wall to the footwall in the general sequence: massive sphalerite, massive pyrite, chalcopyrite - sphalerite and chalcopyrite - pyrrhotite. A chalcopyrite stringer ore and a disseminated pyrite ore occur in the feeder zones. The ores were classified by the company geologists and Healy (Healy and Petruk, 1988) as massive sphalerite (MS), banded pyrite + sphalerite (BP+S), massive pyrite (MP), mixed/banded chalcopyrite + sphalerite (M/BC+S), sheared chalcopyrite + sphalerite (SC+S), massive chalcopyrite + pyrrhotite (MC+P), vein quartz and chalcopyrite (VQ+C), chalcopyrite stringers (CS), and disseminated pyrite and chalcopyrite ores (DP+C). The chemical and mineralogical compositions of the various ore types are given in Tables 4.1 and 4.2.

The Callinan ores are enriched in pyrite and pyrrhotite. Healy classified samples from the Callinan deposit as pyrite, sheared pyrite, pyrrhotite, and sheared pyrrhotite ore types.

Table 4.1.
Chemical compositions of ore types

		Cu (%)	Zn (%)	Pb (%)	Fe (%)	S (%)	As ppm	Ag ppm	Au ppm	Se ppm	Te ppm	Hg ppm	Sn ppm	Sb ppm	In ppm
ORE TYPES		AVERAGE CHEMICAL COMPOSITIONS													
Vein qtz & cp	TL	3.2	0.4	0.01	5.6	3.9	19	13.2	6.17	16	1.5	7	nd	2	nd
Chalcopyrite stringers	TL	12.4	2.7	0.05	27.5	24.0	2800	38.0	2.26	155	2.7	38	38	45	39
Massive cp & po	TL	12.1	0.4	0.01	37.3	29.0	560	39.0	3.36	162	3.7	44	43	40	32
Disseminated cp & py	TL	2.0	0.6	0.02	25.5	21.6	332	8.1	1.27	17	0.5	24	20	43	14
Sheared cp & sp	TL	5.2	23.2	0.01	17.0	22.8	210	65.0	6.86	34	1.1	27	38	71	38
Mixed/banded cp & sp	TL	5.4	17.8	0.01	21.1	25.1	460	58.0	4.05	14	0.8	26	37	69	45
Massive pyrite	TL	0.2	3.3	0.3	38.1	39.5	1080	44.0	1.02	38	0.2	39	10	76	22
Banded py & sp	TL	0.6	19.9	1.1	25.0	33.8	1394	80.0	4.01	27	0.4	79	45	167	35
Massive sphalerite	TL	1.7	37.9	3.1	11.6	25.3	490	131.0	1.51	18	0.4	27	44	148	43
Pyrite	CAL	0.6	2.9	0.13	32.2	39.1	20	18.9	2.74	117	na	32	na	31	na
Sheared pyrite	CAL	1.2	11.4	0.45	30.3	40.1	125	16.1	1.10	55	na	31	na	11	na
Pyrrhotite	CAL	0.7	3.5	0.05	40.4	31.0	40	18.2	1.58	105	na	23	na	3	na
Sheared po	CAL	3.9	7.6	0.07	26.0	27.3	500	7.5	1.3	87	na	28	na	2	na

qtz = quartz; cp = chalcopyrite; po = pyrrhotite; py = pyrite; sp = sphalerite; nd = not detected; na = not analysed
TL = Trout Lake ore; CAL = Callinan ore.

Table 4.2.
Mineral quantities in ore types.

Minerals	VQ+C	CS	MC+P	DC+P	SC+S	M/BC+S	MP	BP+S	MS	Py	S py	Po	S po
Chalcopyrite	9.4	36.5	35.7	3.9	15.4	16.0	0.6	1.8	5.1	1.7	3.6	1.9	10.3
Sphalerite	0.9	4.6	0.7	0.6	38.7	29.6	5.6	33.2	63.2	5.0	20.2	5.9	11.9
Galena	0.0	0.05	0.0	0.01	0.01	0.01	0.3	1.2	3.6	0.1	0.3	0.03	0.04
Pyrite	2.1	20.3	21.3	38.5	3.3	16.3	74.8	42.4	5.1	54.6	48.1	6.9	13.6
Arsenopyrite	0.0	0.5	0.07	0.05	0.04	0.08	0.25	3.24	0.09	0.003	0.20	0.005	0.21
Pyrrhotite	0.9	5.0	21.3	0.03	11.0	7.7	0.01	0.02	2.7	0.4	1.8	50.6	17.6
Magnetite		X								X	X	X	X
Tetrahedrite					X		X	X	X	X			X
Silver									X	X	X	X	X
Pyrargyrite								X	X				
Dyscrasite									X				
Au-Ag-Hg alloy	X	X	X	X	X	X	X	X	X				

Note: Quantities are reported in wt % and ppm; X denotes that the mineral was found in the ore type.

VQ+C = vein quartz + chalcopyrite TL; CS = chalcopyrite stringers TL; MC+P = massive chalcopyrite + pyrrhotite TL; DC+P = Disseminated chalcopyrite + pyrite TL; SC+S = sheared chalcopyrite + sphalerite TL; M/BC+S = massive/banded chalcopyrite + sphalerite TL; MP = massive pyrite TL; BP+S = banded pyrite + sphalerite TL; MS = massive sphalerite TL; Py = pyrite ore CAL; S py = sheared pyrite ore CAL; Po = pyrrhotite ore CAL; S po = sheared pyrrhotite ore CAL.

Table 4.2.
(Contd), Mineral quantities in ore types.

Minerals	VQ+C	CS	MC+P	DC+P	SC+S	M/BC+S	MP	BP+S	MS	Py	S py	Po	S po
Cubanite	X	X	X	X		X							
Boulangerite								X					
Freieslebenite								X					
Hessite		X								X			
Altaite										X	X		
Rucklidgeite		X											
Pilsenite		X											
Clausthalite		X				X			X				
Gudmundite								X	X				
Gangue	86.8	33.0	20.9	61.5	31.5	30.3	18.4	18.1	20.2	37.9	25.6	34.8	46.4

Note: Quantities are reported in wt %. X denotes that the mineral was found in the ore type; gangue = silicates and carbonates

VQ+C = vein quartz + chalcopyrite TL; CS = chalcopyrite stringers TL; MC+P = massive chalcopyrite + pyrrhotite TL; DC+P = Disseminated chalcopyrite + pyrite TL; SC+S = sheared chalcopyrite + sphalerite TL; M/BC+S = massive/banded chalcopyrite + sphalerite TL; MP = massive pyrite TL; BP+S = banded pyrite + sphalerite TL; MS = massive sphalerite TL; Py = pyrite ore CAL; S py = sheared pyrite ore CAL; Po = pyrrhotite ore CAL; S po = sheared pyrrhotite ore CAL.

4.2.1. Trout Lake ores

4.2.1.1. Massive sphalerite (MS) ore type
The massive sphalerite ore type occurs at the hanging wall of the ore lenses, and consists of semi-massive to massive sphalerite. The sphalerite contains gangue, pyrite and minor to trace amounts of chalcopyrite, pyrrhotite, galena, freibergite, gudmundite, pyrargyrite, arsenopyrite and Au-Ag-Hg alloy as interstitial fillings between the sphalerite grains and as inclusions in sphalerite (Figure 4.1). Some of the chalcopyrite is present as irregular grains, but much occurs as minute blebs and rods in sphalerite, in a texture that is referred to as "chalcopyrite disease" (Barton and Bethke, 1987).

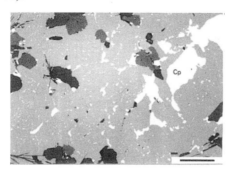

Figure 4.1. Massive sphalerite (MS) ore type showing massive sphalerite (grey) with gangue (black) inclusions and chalcopyrite (white, cp) in interstices between sphalerite grains and as "chalcopyrite disease". Bar scale = 325 μm.

4.2.1.2. Banded pyrite + sphalerite (BP+S) ore type
The banded pyrite + sphalerite ore type occupies an intermediate zone between the massive sphalerite and massive pyrite in the upper parts of the ore lenses. The ore type consists of pyrite and sphalerite bands that range from less than 1 mm to 1 cm in width, and of diffuse laminae of gangue (Figure 4.2 (a)). The pyrite bands consist of massive pyrite, and contain porphyroblasts of euhedral pyrite, arsenopyrite and magnetite, and minute grains of Au-Ag-Hg alloy along pyrite grain boundaries and fractures. The massive pyrite contains sphalerite, chalcopyrite, galena and pyrrhotite in the interstices. The sphalerite bands consist of massive sphalerite and contain chalcopyrite, galena and pyrrhotite in interstices between sphalerite grains. Some sphalerite grains contain fine blebs of chalcopyrite as the "chalcopyrite disease". Other minerals observed in this ore type are freibergite, gudmundite, pyrargyrite, freieslebenite, boulangerite and dyscrasite associated with galena (Figure 4.2 (b)).

4.2.1.3. Massive pyrite (MP) ore type
The massive pyrite ore type occurs in the upper to central parts of the ore lenses. It consists of semi-massive to massive pyrite with interstitial sphalerite, chalcopyrite, and minor galena, freibergite and pyrrhotite (Figure 4.3). The pyrite occurs as masses and as discrete euhedral pyrite porphyroblasts up to 500 μm in diameter. Commonly the pyrite has been fractured and the fractures are filled with other minerals, particularly chalcopyrite. Magnetite is present as subhedral to anhedral grains, and arsenopyrite and Au-Ag-Hg alloy have been observed.

Figure 4.2. Banded pyrite + sphalerite (BP+S) ore type showing:

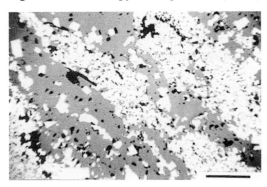

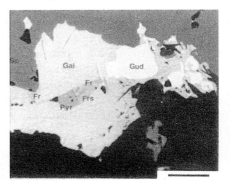

a). Sphalerite (grey) and pyrite (white) bands. Bar scale = 325 µm.

b). Massive galena (Gal) in sphalerite (dark grey band above galena). The galena contains inclusions of gudmundite (Gud), freibergite (Fr), pyrargyrite (Pyr), and freieslebenite (Frs). Bar scale = 40 µm.

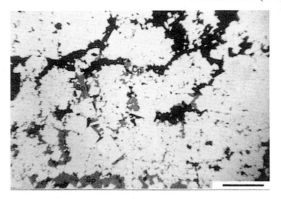

Figure 4.3. Massive pyrite (MP) ore type showing massive pyrite (light grey) with interstitial gangue (black) and sphalerite (grey). Bar scale = 325 µm.

4.2.1.4. Mixed/banded chalcopyrite + sphalerite (M/BC+S) ore type

The mixed/banded chalcopyrite + sphalerite ore type occurs across the entire width of the ore lenses, except in the feeder zone. The ore consists of massive to semi-massive sulfides with inclusions of gangue. The main sulfides are chalcopyrite, sphalerite and pyrrhotite, and they contain rounded grains and irregular masses of pyrite (Figure 4.4). The pyrite masses contain reticulate-textured magnetite, but some magnetite also occurs as disseminated grains and clusters of grains in chalcopyrite and sphalerite. Minor amounts of arsenopyrite, clausthalite and Au-Ag-Hg alloy were found.

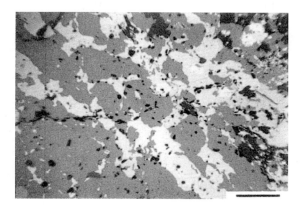

Figure 4.4. Mixed/banded chalcopyrite + sphalerite (M/BC+S) ore type showing sphalerite (dark grey), chalcopyrite (grey) and gangue (black). A minor amount of pyrite (light grey) is present, but is barely distinguishable from chalcopyrite, in the photomicrograph. Bar scale = 325 μm.

4.2.1.5. Sheared chalcopyrite + sphalerite (SC+S) ore type

The sheared chalcopyrite + sphalerite ore type contains more than 5 % of both chalcopyrite and sphalerite, and occurs mainly in the central portions of the ore lenses. It is designated as a metamorphic rather than a stratigraphic ore type, as it displays many textures of a metamorphosed base metal ore (Healy and Petruk, 1988). The metamorphic and deformational textures displayed by this ore type are:

- Chalcopyrite and sphalerite are intimately intergrown and interpenetrate in a texture referred to as "diablastic" (Figure 4.5 (a), (b)), which is interpreted as a deformational texture.
- The chalcopyrite, sphalerite and pyrrhotite commonly display a foliated sulfide matrix and in places the minerals are very fine-grained (Figure 4.5 (a)).
- The gangue in the massive sulfides occurs primarily as augens (eyes) (Figure 4.5 (c), (d)).
- The ore type contains a significant amount of pyrrhotite which may be secondary after pyrite.
- Some of the pyrite is present as rounded relicts in the sphalerite, chalcopyrite and pyrrhotite (Figure 4.5 (e)).
- Some pyrite is intensely fractured and the fractures are filled with chalcopyrite and minor pyrrhotite (Figure 4.5 (f)).
- Some pyrite contains the reticulate-textured magnetite.
- The chalcopyrite commonly contains exsolution lamellae of cubanite.

Trace amounts of galena, gudmundite and Au-Ag-Hg alloy were found in this ore type.

4.2.1.6. Massive chalcopyrite + pyrrhotite (MC+P) ore type

The massive chalcopyrite + pyrrhotite ore type occurs at the footwall of the massive sulfide ore and extends towards the centre. The MC+P ore type consists of massive to semi-massive sulfides that are composed of relatively wide bands of chalcopyrite-pyrrhotite heterogeneously interlayered with narrower bands of chalcopyrite-sphalerite and magnetite-pyrite. The pyrite and chalcopyrite display obvious evidence of metamorphism. In particular, some of the pyrite occurs as fractured and rounded porphyroblasts, some occurs as relict pyrite in chalcopyrite, pyrrhotite and sphalerite, and some contains reticulate-textured magnetite. The chalcopyrite commonly contains slightly bent exsolution lamellae of cubanite.

80

Figure 4.5. Sheared chalcopyrite + sphalerite (SC+S) ore type showing:

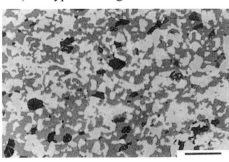

a). Diablastic textured chalcopyrite (white) and sphalerite (grey). Bar scale = 160 µm.

b). Diablastic textured chalcopyrite and sphalerite. Bar scale = 40 µm.

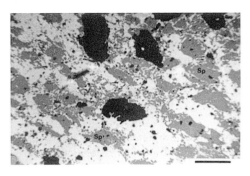

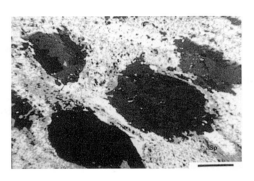

c). Augen textured gangue (black) and elongated sphalerite (grey) and chalcopyrite (white). Bar scale = 325 µm.

d). Augen textured gangue (black) in very fine grained diablastic chalcopyrite (white) and sphalerite (grey). Bar scale = 325 µm.

e). Rounded pyrite porphyroblasts (white, Py) in sheared sphalerite-chalcopyrite-pyrrhotite. Bar scale = 325 µm.

f). Fractured pyrite (white) porphyroblast with chalcopyrite and minor pyrrhotite (light grey) filling fractures. Bar scale = 160 µm.

4.2.1.7. Vein quartz and chalcopyrite (VQ+C) ore type

The vein quartz and chalcopyrite ore type occurs as mineralised quartz veins within the ore lenses, without stratigraphic control, and may have resulted by hydrothermal remobilisation during metamorphism (Koo and Mossman, 1975). The veins consist of coarse-grained quartz with minor calcite, sericite, chlorite and tourmaline, and contain coarse-grained chalcopyrite in the interstitial spaces between the quartz. Small amounts of sphalerite, cubanite, pyrite, pyrrhotite, marcasite and galena are also present.

4.2.1.8. Chalcopyrite stringer (CS) ore type

The chalcopyrite stringer ore type, together with the disseminated pyrite + chalcopyrite ore type underlie the massive sulfides as a footwall stockwork mineralization and represent a feeder zone. The chalcopyrite stringer ore consists of semi-massive disseminated sulfides in a foliated chloritic schist (Figure 4.6 (a)(b)). The sulfides are coarse-grained chalcopyrite with smaller amounts of pyrrhotite, sphalerite, and pyrite. Other sulfide minerals are cubanite, galena, clausthalite, rucklidgeite, Au-Ag-Hg alloy, hessite, pilsenite and magnetite. Some of the pyrite occurs as masses, but much of it displays evidence of metamorphism. In particular, some pyrite occurs as porphyroblasts in gangue, some occurs as corroded and rounded grains in chalcopyrite, and some contains fine-grained reticulate-textured magnetite.

Figure 4.6. Chalcopyrite stringer (CS) ore type showing:

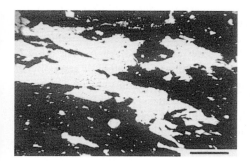 

a). Chalcopyrite and minor pyrite and pyrrhotite (white, not differentiated) in gangue (black). Bar scale = 325 μm.

b). Chalcopyrite and minor pyrite and pyrrhotite (white, not differentiated) in chlorite (dark grey). Bar scale = 325 μm.

4.2.1.9. Disseminated pyrite and chalcopyrite (DP+C) ore type

The disseminated pyrite + chalcopyrite ore type occurs as a diffuse envelope around the chalcopyrite stringer ore type in the footwall stockwork mineralization. It consists of disseminated euhedral pyrite in gangue (Fig 4.7), and contains small amounts of chalcopyrite, pyrrhotite, sphalerite, marcasite, magnetite, galena, clausthalite and Au-Ag-Hg alloy.

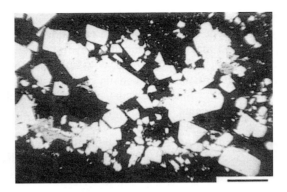

Figure 4.7. Disseminated pyrite + chalcopyrite (DP+C) ore type showing disseminated pyrite (white) and minor chalcopyrite (light grey) in gangue (black). Bar scale = 325 μm.

4.2.2 Callinan ore types

4.2.2.1 Pyrite ore type

The pyrite ore type consists largely of massive to semi-massive pyrite, but locally, the pyrite is interlayered with sphalerite. The massive pyrite is composed of individual pyrite grains that are commonly foliated, granulated and fractured (Figure 4.8 (a)). The pyrite contains sphalerite, chalcopyrite, gangue, pyrrhotite, galena and altaite in interstitial spaces between the pyrite grains and clusters (Figure 4.8 (b), 4.8 (c)), and galena, pyrrhotite, altaite and Au-Ag-Hg alloy in fractures. Small masses of chalcopyrite and sphalerite, and grains and clusters of arsenopyrite are also present in the massive pyrite. The sphalerite layers consist largely of a massive sphalerite that contains pyrite and chalcopyrite. The pyrite in the massive sphalerite is present as large porphyroblasts and as disseminated grains that exhibit atoll textures. The chalcopyrite in the massive sphalerite is present as small masses and as the so-called "chalcopyrite disease". The grain size distributions of the pyrite, sphalerite and chalcopyrite are 80% minus 110 μm, 110 μm and 39 μm, respectively.

Rare and extremely coarse-grained (up to 3 mm) magnetite porphyroblasts are present in the ore. The magnetite generally contains fine-grained (<25 μm) inclusions of pyrite and small amounts of chalcopyrite, sphalerite, pyrrhotite, and Au-Ag-Hg alloy.

4.2.2.2. Sheared pyrite ore

The sheared pyrite represents a section of high grade ore, and consists essentially of sheared banded pyrite and sphalerite. The texture ranges from massive pyrite with chalcopyrite, sphalerite and other minerals in interstitial spaces (Figure 4-9 (a)), through pyrite porphyroblasts and atolls in massive sphalerite, to rounded pyrite in massive sphalerite (Figure 4.9 (b)). Intimate intergrowths of chalcopyrite and sphalerite displaying a diablastic texture are also present (Figure 4.9 (c)). The grain size distributions of the pyrite, sphalerite and chalcopyrite are 80% minus 130 μm, 110 μm and 40 μm, respectively.

Figure 4.8. Textures of pyrite ore type, Callinan deposit:

a). Massive pyrite (white) with gangue (black) and sphalerite (dark grey) in interstitial spaces. Bar scale = 160 μm.

b). Semi-massive pyrite (white) with gangue (black) and sphalerite (dark grey) in interstitial spaces. Bar scale = 160 μm.

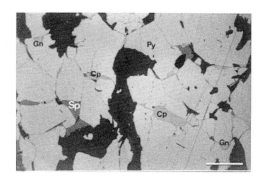

c). Massive pyrite (white, Py) with chalcopyrite (light grey, Cp), sphalerite (dark grey, Sp) and gangue (black) in interstices. Bar scale = 160 μm.

4.2.2.3. Pyrrhotite ore type

The pyrrhotite ore type is a semi-massive to massive sulfide ore composed of pyrrhotite, sphalerite and chalcopyrite and contains porphyroblasts of pyrite and magnetite (Figure 4.10 (a), 4.10 (b), 4.10 (c), 4.10 (d)). The grain size distributions of the sphalerite and chalcopyrite are 80% minus 130 μm, 110 μm, respectively.

4.2.2.4. Sheared pyrrhotite ore type

The sheared pyrrhotite ore type represents a high grade section of massive to semi-massive sulfide ore. It is composed of massive pyrrhotite, sphalerite and chalcopyrite and contains porphyroblasts of pyrite and magnetite. The ore is highly disrupted and foliated, and displays many features of a sheared and metamorphosed ore including diablastic textured sphalerite and chalcopyrite. The grain size distributions of the sphalerite and chalcopyrite are 80% minus 140 μm, 200 μm, respectively.

84

Figure 4.9. Textures of minerals in sheared pyrite ore type, Callinan deposit:

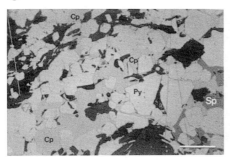

a). Massive pyrite (white, Py) with chalcopyrite (light grey, Cp), sphalerite (grey, Sp) and gangue (black) in interstices. Bar scale = 160 μm.

b). Rounded pyrite grains (white, Py) in massive sphalerite (grey). The sphalerite contains minute grains of chalcopyrite (Cp) and pyrrhotite (Po). Bar scale = 160 μm.

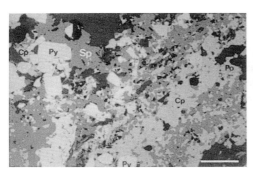

c). Diablastic texture of sphalerite (grey, Sp) and chalcopyrite (light grey, Cp), with minor pyrite (Py and pyrrhotite (Po). Bar scale = 160 μm.

Figure 4.10. Textures of minerals in pyrrhotite ore type, Callinan deposit:

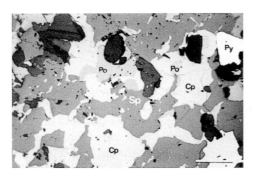

a). Intergrowth of massive sphalerite (grey, Sp), chalcopyrite (white, Cp) and pyrrhotite (light grey, Po) with an inclusion of pyrite (white, Py). Bar scale = 160 μm.

b). Intergrowth of massive sphalerite (grey) and pyrrhotite (light grey). Bar scale = 160 μm.

Figure 4.10. (Cont'd) Textures of minerals in pyrrhotite ore type, Callinan deposit:

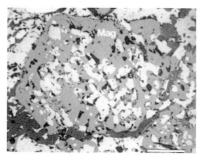

c). Massive pyrrhotite (light grey, Po) with pyrite porphyroblasts (white, Py) that contain trapped sphalerite. Bar scale = 160 μm.

d). Coarse-grained magnetite porphyroblast (grey, mag) with inclusions of chalcopyrite and pyrrhotite (white). Bar scale = 160 μm.

4.3. MINERALS IN TROUT LAKE AND CALLINAN ORES

4.3.1. Pyrite (FeS$_2$)

Pyrite is the most abundant mineral in the Trout Lake and Callinan ores. It is present in all ore types, and displays characteristics of regional and tectonic metamorphism. It occurs as (1) polycrystalline masses (Figure 4.3), (2) rounded inclusions (commonly porphyroblasts) in sphalerite and chalcopyrite (Figure 4.5 (e)), (3) fractured porphyroblasts (Figure 4.5 (f)), (4) disseminated porphyroblasts and metacrysts in gangue (Figure 4.11 (a)), (5) schistose and deformed pyrite (Figure 4.11 (b)), (6) atoll texture which exhibits embayments of sphalerite in pyrite (Figure 4.11 (c)), (7) replacement by sphalerite, chalcopyrite or galena (Figure 4.11 (d)), (8) reticulate textured magnetite in pyrite (Figure 4.11 (e), (9) replacement by magnetite (Figure 4.11 (f)), (10) replacement by marcasite in zones of intense deformation (e.g. along shear zones) and (11) remobilised pyrite in veinlets in gangue minerals. The massive pyrite contains sphalerite, gangue and chalcopyrite in interstitial spaces, and veinlets of galena and a variety of other minor minerals.

4.3.2. Pyrrhotite (Fe$_{(1-x)}$S)

Pyrrhotite is an abundant mineral in the Trout Lake deposit, and is the major constituent in the massive chalcopyrite + pyrrhotite (MC+P) ore type. The mineral is commonly associated with chalcopyrite and tends to be concentrated in chalcopyrite-rich ore types. The pyrrhotite content in the sheared chalcopyrite + sphalerite (SC+S) ore type tends to increase with intensity of deformation and a decrease in pyrite content, which suggests that the pyrrhotite in the SC+S ore type formed by a breakdown of pyrite during deformation. Most of the pyrrhotite is monoclinic and moderately magnetic.

Pyrrhotite is a major constituent of the pyrrhotite and sheared pyrrhotite ores in the Callinan deposit where it occurs as large grains enveloping disseminated pyrite and magnetite porphyroblasts, and as remnants in the magnetite porphyroblasts.

Figure 4.11. Pyrite textures:

a). Euhedral pyrite (white) in sphalerite (grey) and gangue (black). Bar scale = 160 μm.

b). Deformed and schistose pyrite. Bar scale = 325 μm.

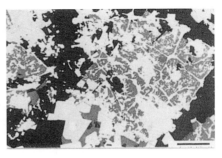

c). Euhedral pyrite (white) in sphalerite (dark grey); displaying atoll texture. Bar scale = 325 μm.

d). Dendritic sphalerite (grey) replacing pyrite (white). Bar scale = 160 μm.

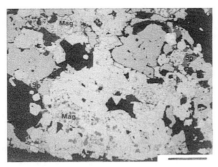

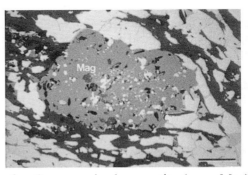

e). Massive pyrite (light grey) with chalcopyrite and pyrrhotite (grey, undifferentiated), and gangue (black). The pyrite in the lower central part of photomicrograph contains fine-grained reticulate textured magnetite (Mag). Bar scale = 160 μm.

f). Coarse-grained magnetite (grey, Mag) porphyroblast in pyrite (white), containing pyrite inclusions. Callinan ore. Bar scale = 160 μm.

4.3.3. Chalcopyrite (CuFeS$_2$)

Chalcopyrite is the main Cu-bearing mineral in the Trout Lake and Callinan ores. It is abundant near the footwall and central parts of the massive sulfides, and in the underlying stringer ore. The grain size of the mineral varies from coarse-grained (up to 1 cm in diameter) to fine-grained (1 to 5 μm blebs and lamellae in sphalerite), with grain size decreasing as the abundance of chalcopyrite decreases. In particular, size distribution data shows that the chalcopyrite varies from very coarse-grained (up to 1 cm in diameter) in the vein quartz +chalcopyrite (VQ+C) ores, to 80 % minus 212 μm in massive chalcopyrite + pyrrhotite (MC+P) and chalcopyrite stringer (CS) ores, to 80 % minus 13 μm in massive pyrite (MP) ore. It is estimated that about 80 % of the chalcopyrite in the Trout Lake deposit is coarse-grained (e.g. ~80 % minus 150 μm). Chalcopyrite exhibits the characteristics of a metamorphosed ore which include (1) intimate intergrowths (diablastic texture) with sphalerite and/or pyrrhotite (Figure 4.5a and 4.5b), (2) massive coarse-grained chalcopyrite that invades, replaces and envelopes all minerals including gangue (Figure 4.4), (3) massive coarse-grained chalcopyrite that contains cubanite lamellae (Figure 4.12 (a)), (4) massive coarse-grained chalcopyrite that contains sphalerite stars (Figure 4.12 (b)) and (5) veinlets in pyrite and replacements of pyrite (Figure 4.5 (f)). The chalcopyrite also occurs in interstices between sphalerite grains, and as "chalcopyrite disease" in sphalerite (Figure 4.12 (c)).

Figure 4.12. Chalcopyrite textures:

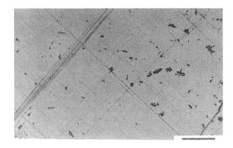

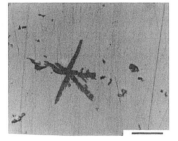

a). Cubanite lamellae (narrow dark grey lamellar mineral at left of photomicrograph) in massive chalcopyrite (grey). Bar scale = 160 μm.

b). Sphalerite star (dark grey) in massive chalcopyrite (grey). Bar scale = 160 μm.

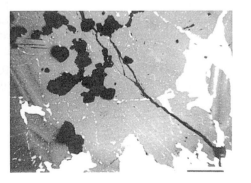

c). Large chalcopyrite grains and minute chalcopyrite blebs and lamellae (white) aligned along crystallographic planes in sphalerite (grey). Bar scale = 325 μm.

4.3.4. Sphalerite (ZnS)

Sphalerite is ubiquitous, and was observed in all samples examined. It occurs as a major mineral near the hangingwall and central parts of the massive sulfides, and as a minor to trace mineral near the footwall. The grain size of the mineral varies from coarse-grained (≤ 3 mm) to fine-grained, with grains size decreasing as the abundance of sphalerite decreases. Size distribution data show that it varies from 80 % minus 300 μm in the massive sphalerite (MS) ores to 80 % minus 52 μm in massive pyrite (MP) ores. It is estimated that the size distribution of most of the sphalerite in the Trout Lake deposit is about 80 % minus 75 μm. The sphalerite occurs as (1) masses with small inclusions of the other minerals in the ore, (2) interstitial fillings in pyrite and replacements of pyrite (including atoll textured pyrite), and (3) intergrowths with chalcopyrite (including diablastic texture). The Fe content in sphalerite varies from 1.7 to 8.4 wt % with a mean of 7.0 wt %, Zn content varies from 57.1 to 64.9 wt % with a mean of 59.2 wt %, and the Cd content varies from non-detectable to 0.6 wt % with mean of 0.3 wt %.

4.3.5. Galena (PbS)

Galena is generally a trace constituent in the ores, although occasionally it is a minor, and rarely a major constituent. It occurs primarily near the hanging wall in the massive sphalerite (MS) and banded sphalerite +pyrite (BS+P) ore types, and is a major constituent only in sphalerite-rich ores. The galena occurs as (1) irregular grains in massive sphalerite (Figure 4.13 (a)), (2) interstitial fillings, psuedoveinlets and small masses in massive sphalerite and massive pyrite (Figure 4.13 (b)), (3) dendritic galena replacing pyrite (Figure 4.13 (c)), and (4) disseminations in chalcopyrite, sphalerite and pyrrhotite. The galena occurring as interstitial fillings and psuedoveinlets is commonly associated with freibergite, pyrargyrite, gudmundite, boulangerite, freieslebenite, bournonite, dyscrasite and Au-Ag-Hg alloy. Size distribution data give a cumulative grain size of 80 % minus 52 μm. Some of the galena occurs as a solid solution with clausthalite (Healy and Petruk, 1992). Microprobe analyses detected trace amounts of Fe, Te and reported a maximum of 1,200 ppm Ag and a mean of 660 ppm Ag in the Trout Lake galena (Pinard and Petruk, 1989).

4.3.6. Au-Ag-Hg alloy

The term Au-Ag-Hg alloy is used here instead of native gold or electrum because the composition of the mineral varies from 75 % Au, 22.2 % Ag, 1.9 % Hg through 20.8 % Au, 68.2 % Ag, 12.3 % Hg, to 1.6 % Au, 66,3 % Ag, 30.9 % Hg (Healy and Petruk, 1990). The Au-Ag-Hg alloy occurs in all ore types. It is present as (1) fracture fillings and inclusions in massive pyrite and films on pyrite grains (Figure 4.14 (a), 4.14 (b)), (2) inhomogeneous masses and veinlets at the contact between chalcopyrite and chlorite in footwall ores (Figure 4.14 (c)), and (3) fine-grained inclusions in galena/freibergite veinlets that occur as fracture fillings in pyrite (Figure 4.14 (b)), and in interstices between pyrite grains (Healy and Petruk, 1990d).

Figure 4.13. Textures of galena:

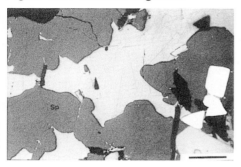

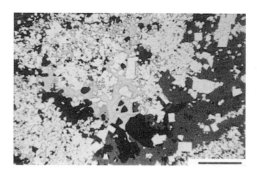

a). Massive sphalerite (grey) with an interstitial mass of galena (light grey), and a few grains of pyrite (white). Trace amounts of gudmundite and freibergite (not differentiated) are associated with the galena. Bar scale = 160 μm.

b) Massive galena (light grey) replacing pyrite (white). Bar scale = 325 μm.

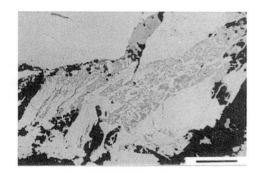

e) Dendritic galena (grey) replacing pyrite (light grey) along a shear zone. Bar scale = 160 μm.

4.3.7. Other Minerals

Other minerals found in the Trout Lake ore are cubanite ($CuFe_2S_3$), arsenopyrite (FeAsS), gudmundite (FeSbS), magnetite (Fe_3O_4), galena (PbS), acanthite (Ag_2S), dyscrasite (Ag_3Sb), freibergite ($(Cu,Ag,Fe)_{12}Sb_4S_{13}$), pyrargyrite ($Ag_3SbS_3$), freieslebenite ($PbAgSbS_3$), native silver (Ag), hessite (Ag_2Te), volynskite ($BiAgTe_2$), pilsenite ($(Bi,Pb)_8(Te,Se)_6$), rucklidgeite ($(Bi,Pb)_3Te_4$), boulangerite ($Pb_5Sb_4S_{11}$), bournonite ($PbCuSbS_3$), costibite (CoSbS), clausthalite (PbSe), naumannite (Ag_2Se), cassiterite (SnO_2), monazite ($(Ce,La,Nd,Th)PO_4$) and bastnasite ($(Y,Ce)(CO_3)F$). In addition, altaite (PbTe) was found in the Callinan deposit.

Sixty analyses of friebergite grains were acquired with a microprobe, even though friebergite is not the major Ag-bearing mineral in the ore. The mean composition in wt % is 21.6 Cu, 21.7 Ag, 26.7 Sb, 0.2 As, 22.8 S, 5.6 Fe, and 1.4 Zn. The minimum and maximum amounts of the respective elements, in wt %, are 14.7 - 28.5 Cu, 13.0 - 30.3 Ag, 23.6 - 28.1 Sb, 0.0 - 2.8 As, 21.8 - 24.2 S, 4.8 - 6.6 Fe, and 0.0 - 2.6 Zn (Healy and Petruk, 1988).

Figure 4.14. Textures of Au-Ag-Hg alloy:

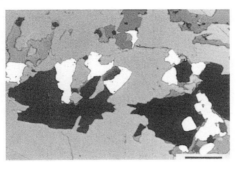

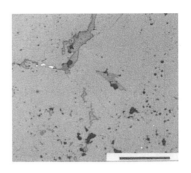

a). Irregular Au-Ag-Hg alloy grains (white) in pyrite (very light grey) associated with galena (grey) freibergite (dark grey) and gangue (black). Bar scale = 40 μm.

b). Very minute grain of Au-Ag-Hg alloy (white) in a galena veinlet (grey) in pyrite (light grey). Bar scale = 160 μm.

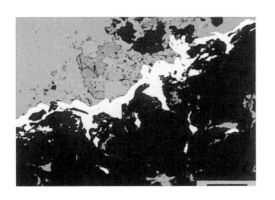

c) Veinlet of Au-Ag-Hg alloy (white) along the contact between chlorite (black) and chalcopyrite (light grey) in a chalcopyrite stringer (CS) ore type. Bar scale = 106 μm.

4.4. MINERAL PROCESSING

A suite of samples was collected from the Trout Lake concentrator during a two hour sampling campaign on Feb. 5. 1986. The flowsheet is given in Figure 4.15. The samples were assayed for Cu, Zn, Pb, Fe, S, Au, Ag, As, Sb, Ni, Cd, Co, Mn, Te, Se, Sn, In and Hg, and the mineral quantities were determined by a combination of image analysis and calculations from chemical assays. A materials balance calculation was performed to determine the flowrates of each product through the concentrator, and to determine the recoveries (distributions) of the elements and minerals in each product. Values for the feed, rougher concentrates, final concentrates and tails are given in Tables 4.4 and 4.5 (Healy and Petruk, 1989a; 1990e).

The feed to the concentrator was as a composite of all ore types plus dilution by wallrock. A materials balance (MATBAL5) (Laguitton, 1985) calculation was performed to determine the proportion of each ore type, and the contribution of Cu, Zn, Fe, S, Pb, Au, Ag, pyrite, pyrrhotite and gangue by each ore type (Table 4.3).

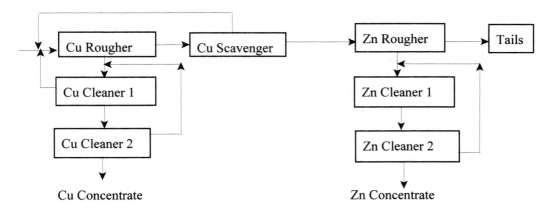

Figure 4.15. Generalized flowsheet of the Trout Lake concentrator, February, 1986.

4.4.1. Cu concentrate

Tables 4.4 and 4.5 show a high recovery of Cu /chalcopyrite (91.7 %) in a high grade copper concentrate (28.0 wt % Cu). The Cu concentrate also recovered 59.2 % of the Au, 48.5 % of the Ag, and 13.2 % of the Pb. It contains 5.9 % of the pyrite, 10.9 % of the Zn, 80.6 % of the Te, 26.5 % of the Se, and a notably low proportion of the Sb (18.4 %).

4.4.1.1. Chalcopyrite

The high Cu recovery was obtained because the chalcopyrite is relatively coarse grained in the ore types that contained most of the Cu (chalcopyrite stringer, massive/banded chalcopyrite + sphalerite, sheared chalcopyrite + sphalerite and massive chalcopyrite + pyrrhotite). The flotation feed had been ground to about 83 % minus 75 μm (minus 200 mesh), and the Cu rougher tails was reground and scavenged. The scavenger concentrate was recirculated to the flotation rougher cell (Figure 4.15).

About 62 % of the chalcopyrite in the circuit was totally liberated, and 94 % was in particles that contain more than 70 % chalcopyrite. About 99 % of the totally liberated chalcopyrite, and 80 % of the unliberated chalcopyrite (mostly in 70 to 99.9 % particles) in the feed was recovered in the Cu concentrate. About 75 % of the chalcopyrite in the Cu concentrate was liberated. The remaining 25 % was in particles that contained gangue, pyrite, pyrrhotite and sphalerite. In particular, about 13 % of the chalcopyrite in the Cu concentrate was in particles with gangue as the main impurity, about 8 % was in particles with pyrite and pyrrhotite as the main impurities, and about 4 % was in particles with sphalerite as the main impurity. The size distribution of the liberated chalcopyrite grains or particles recovered in the Cu concentrate was about 84 % minus 26 μm with most grains or particles being between 6.5 and 26 μm. The size distribution of unliberated chalcopyrite grains within larger host particles was about 85 % minus 37.5 μm with most chalcopyrite grains being between 13 and 37 μm.

Table 4.3.
Assays of feed, and contributions to feed by each ore type

PRODUCT	Wt % of product	Cu wt %	Zn wt %	Pb wt %	Fe wt %	S wt %	Au ppm	Ag ppm	Pyrite wt %	Gangue wt %	Pyrrhotite wt %
Feed (content, wt %)	100	2.16	2.57	0.038	16.02	9.61	2.23	12.35	10.6	77.6	0.97
Ore type	Contributions to feed by each ore type , as percent of element or mineral in feed.										
Massive sp	0.3	0.2	4.2	23.2	0.2	0.8	0.2	3.0	.05	0.1	0.8
Banded py+sp	2.2	0.5	13.4	55.4	3.7	7.7	4.2	14.1	10.3	0.5	0.02
Massive py	9.6	1.3	15.5	2.0	21.4	36.2	6.8	6.8	59.9	2.0	0.2
Massive/Banded cp+sp	2.8	6.8	18.5	0.5	3.6	7.1	6.5	12.2	3.6	1.1	20.5
Sheared cp+sp	2.3	5.4	19.7	7.2	2.4	5.3	10.0	11.3	1.1	0.9	23.9
Massive cp+po	0.3	1.6	0.05	0.13	0.7	0.9	0.5	0.9	0.5	0.1	6.3
Chalcopyrite stringer	27.4	83.4	26.8	11.5	38.6	39.5	71.0	51.2	21.5	16.1	48.2
Disseminated cp+py	1.0	0.6	1.8	0.03	1.0	2.5	0.5	0.5	3.0	0.8	0.03
Vein qtz+cp	0.1	0.2	0.05	0.01	0.04	0.04	0.3	0.01	0.03	0.1	0.1
Wallrock	54.0	0.0	0.0	0.0	28.4	0.0	0.0	0.0	0.0	78.3	0.0
Total	100.0	100.0	100.0	100.0	100.0	100.0	100.0	100.0	100.0	100.0	100.0

Py = pyrite; sp = sphalerite; cp = chalcopyrite; po = pyrrhotite; qtz = quartz.

Table 4.4.
Assays and element recoveries/distributions for selected Trout Lake products.

ORE TYPES

	Cu (%)	Zn (%)	Pb (%)	Fe (%)	S (%)	As (%)	Ag ppm	Au ppm	Se ppm	Te ppm	Hg ppm	Sn ppm	Sb ppm	In ppm
								ASSAYS						
Feed	2.2	2.6	0.04	16.0	9.6	0.08	12.3	2.23	49.0	0.4	19.0	7.0	34.0	8.0
Copper rougher Conc.	17.4	5.4	0.09	28.1	31.5	0.16	73.8	19.6	147.0	4.6	43.0	55.0	94.0	45.0
Copper Conc.	28.0	4.0	0.07	29.7	34.8	0.04	84.9	18.7	184.0	4.5	40.0	65.0	89.0	51.0
Zinc rougher Conc.	1.4	34.3	0.36	17.0	30.1	0.32	51.2	6.4	85.0	0.1	79.0	35.0	167.0	42.0
Zinc Conc.	1.7	47.3	0.38	12.6	32.9	0.28	59.9	7.6	81.0	0.1	96.0	36.0	198.0	53.0
Circuit Tails	0.12	0.19	0.02	15.0	6.4	0.07	4.1	0.65	37.0	0.1	14.0	1.0	21.0	2.0
						RECOVERIES/DISTRIBUTION (%)								
Feed	100	100	100	100	100	100	100	100	100	100	100	100	100	100
Copper rougher Conc.	108	26.8	32.0	22.2	41.9	26.6	76.0	112	38.1	147	28.6	99.8	35.2	71.4
Copper Conc.	91.7	10.9	13.1	13.2	25.7	3.6	48.5	59.2	26.5	80.6	14.5	65.5	18.4	45.1
Zinc rougher Conc.	4.2	84.2	60.4	6.7	19.8	27.1	26.1	18.0	10.8	1.7	25.8	31.2	31.0.	32.6
Zinc Conc.	3.5	82.5	45.5	3.5	15.4	17.1	21.7	15.3	7.3	0.9	22.3	23.2	26.2	29.6
Circuit Tails	4.8	6.6	41.4	83.3	58.9	79.3	29.8	25.5	66.2	18.5	63.2	11.3	55.4	25.3

Table 4.5.
Mineral quantities and recoveries/distributions for selected Trout Lake products.

PRODUCTS	MINERAL QUANTITIES (wt %)							
	cp	sp	ga	py	po	asp	G	Weight % of product
Feed	6.3	4.4	0.04	10.6	1.0	0.03	78.2	100.0
Copper rougher Conc.	50.2	9.1	0.11	20.5	0.5	0.2	20.8	12.7
Copper Conc.	80.8	6.8	0.08	8.8	0.6	0.07	5.5	7.1
Zinc rougher Conc.	4.0	57.9	0.4	17.3	2.0	0.35	18.3	6.3
Zinc Conc.	4.9	79.9	0.42	8.5	2.0	0.30	5.3	4.5
Circuit Tails	0.35	0.32	0.02	10.8	0.9	0.01	87.7	88.4
	RECOVERIES/DISTRIBUTION (%)							
Feed	100.0	100.0	100.0	100.0	100.0	100.0	100.0	100.0
Copper rougher Conc.	103	26.8	32.1	24.6	7.2	87.5	3.4	12.7
Copper Conc.	91.7	10.9	12.8	5.9	4.3	16.2	0.5	7.1
Zinc rougher Conc.	4.2	84.1	59.8	10.3	12.9	76.1	1.5	6.3
Zinc Conc.	3.5	82.4	45.0	3.6	9.4	46.9	0.3	4.5
Circuit Tails	4.8	6.6	42.2	90.5	86.3	36.9	99.2	88.4

cp = chalcopyrite; sp = sphalerite; ga = galena; py = pyrite; po = pyrrhotite; asp = arsenopyrite; G = gangue.

4.4.1.2. Sphalerite

The Cu rougher concentrate contained 5.4 wt % Zn which accounted for 26.8 % of the sphalerite in the circuit feed, and the Cu concentrate contained 4 wt % Zn which accounted for 10.9 % of the Zn in the circuit feed (Table 4.4). An image analysis study showed that the distribution of sphalerite in the Cu concentrate was (1) 5.4 % apparently liberated sphalerite, (2) 4.3 % unliberated sphalerite in particles containing 70 to 99.9 % sphalerite and (3) 1.2 % unliberated sphalerite in particles containing 0.1 to 70 % sphalerite. Qualitative petrographic observations indicated that most of the unliberated sphalerite in the Cu concentrate was attached to chalcopyrite, and much more of the sphalerite in the Cu concentrate than of sphalerite in the feed contained the "chalcopyrite disease". It is interpreted that the chalcopyrite in the unliberated sphalerite particles, and in particles with the "chalcopyrite disease", acted as a catalyst for precipitation of available Cu ions on the sphalerite surfaces of those particles, and consequently for activation and flotation in the Cu circuit. The selection of apparently liberated sphalerite particles for flotation in the Cu circuit is not as obvious. Nevertheless, some of the apparently liberated sphalerite particles were indeed selectively activated for flotation in the Cu circuit, probably by precipitation of available Cu ions as a coating on the particle surfaces. It is possible that the selectivity criteria for activating apparently liberated sphalerite particles was higher Cu contents in the particles, either as minute inclusions, not observed in polished sections, or as Cu in solid solution. In fact microprobe analyses had shown that the mean Cu content in the apparently liberated sphalerite in the Cu concentrate was 10 times as high as the mean Cu content (0.05 wt %) in apparently liberated sphalerite in the circuit feed (Healy and Petruk, 1988). However, a subsequent evaluation of the microprobe data indicated that the high Cu values could be due to secondary fluorescence and not Cu in solid solution in sphalerite (Healy, 2000).

4.4.1.3. Pyrite

Table 4.5 shows that about 24.6 % of the pyrite in the feed floated in the Cu rougher cells, but most was rejected by the Cu cleaner cells, as only 5.9 % of the pyrite in the feed was in the Cu concentrate which contained 8.8 % pyrite. An image analysis study of the pyrite in the Cu concentrate showed that 4.3 % was apparently liberated, about 1.3 % was in particles that are composed largely of pyrite but contain small to significant amounts of chalcopyrite and other minerals, and about 0.3 % was in particles composed of pyrite and small amounts of gangue. This observation indicates that the apparently free pyrite and the pyrite attached to gangue must have been activated (probably by Cu ions) sufficiently to float in the Cu circuits. The size distribution of the apparently free pyrite in the Cu concentrate was 83 % minus 37.5 μm with most between 9 and 37 μm. The size distribution of the unliberated pyrite grains within larger host particles in the Cu concentrate was 70 % minus 53 μm with most between 13 and 53 μm in size.

4.4.1.4. Gold

Due to a circulating load through the Cu scavenger, the amount of gold in the Cu rougher concentrate was 112 % with respect to the gold content in the circuit feed. However, the Cu cleaners rejected a high proportion of the gold, and only 59.2 % of the gold in the circuit feed was recovered in the Cu concentrate. Since the gold occurs as a Au-Ag-Hg alloy that is present as minute grains and veinlets, largely in pyrite, and as masses along chalcopyrite-chlorite contacts, it is interpreted that the larger Au-Ag-Hg alloy grains were liberated and concentrated in the Cu concentrate, whereas the smaller Au-Ag-Hg alloy inclusions and veinlets in pyrite were not liberated and probably formed part of the pyrite particles that were pulled into the Cu rougher cells, but were rejected by the Cu cleaners. Concentrations of free Au-Ag-Hg alloy grains were observed in the Cu concentrate (Figure 4.16), and minute inclusions of Au-Ag-Hg alloy in pyrite were observed in the Cu scavenger tail but not in the final tails. However, cyanidation of the final tails (without regrinding) recovered 50 % of the gold in the final tails. The cyanidation test indicates that at least 50 % of the gold in the tails was present as minute Au-Ag-Hg alloy grains attached to other minerals, probably pyrite, and exposed to cyanide solutions. SIMS analyses have shown that the pyrite in the Trout Lake ore has a average content of 0.7 ppm invisible gold, and the arsenopyrite has an average gold content of 32 ppm invisible gold (Chryssoulis and Cook, 1988). This amount of invisible gold can account for only about 13 to 15 % of the gold in the tails. Since 50 % of the gold in the tails is exposed, 15 % is 'invisible' gold, 35 % must be encapsulated, most probably in pyrite.

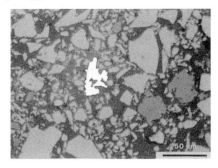

Figure 4.16. Cu concentrate containing chalcopyrite (light grey), a grain of Au-Ag-Hg alloy (white), and a few grains of medium- and fine-grained apparently liberated sphalerite (dark grey). Bar scale = 50 μm.

4.4.1.5. Silver

The behaviour of silver in the concentrator was not characterized, as the silver is hosted by an array of complex silver-bearing minerals, and the grade of Ag in the feed was low (12.3 ppm). About 76 % of the silver in the circuit feed was recovered in the Cu rougher concentrate and 48.5 % in the Cu concentrate at a grade of 84.9 ppm Ag. It is interpreted that most of the recovered silver was a constituent of the Au-Ag-Hg alloy, freibergite and pyrargyrite, which were observed in the Cu concentrate. It is possible that the low Ag recovery was due to poor liberation of the fine-grained Ag-bearing minerals which occur as inclusions and veinlets in pyrite in close association with galena (Figure 4.17).

It is noted that the Ag content in the Trout Lake galena (ave = 660 ppm) is lower than the Ag content in galena from other volcanogenic deposits studied by the senior author. In contrast the Ag content in freibergite is higher (ave. = 21.7 wt %) than in other volcanogenic deposits. These values suggest that some of the Ag was expelled from the Trout Lake galena during metamorphism, and reprecipitated as other Ag-bearing minerals (particularly Ag-rich freibergite) in veinlets associated with galena (Petruk and Wilson, 1993).

4.4.1.6. Galena

Only 32 % of the lead in the feed was recovered in the Cu rougher concentrate, and 13.1 % in the Cu concentrate, which is unusually low. However, the ore contains only 0.04 wt % Pb, and the apparent galena liberation is only about 7 % in the feed and about 18 % in the Cu concentrate. The galena in the ore is relatively fine grained; the apparently free galena grains in the Cu concentrate are smaller than 18.7 µm with most being between 4.7 and 18.7 µm, and the unliberated galena grains are smaller than 26.5 µm with most being between 3.5 and 26.5 µm. It is noteworthy that the galena in this ore contains variable amounts of Se and forms a solid solution with clausthalite (PbSe) (Healy and Petruk 1992). It is possible that the Se rich galena does not float as readily in Cu flotation cells as normal galena.

Fig. 4.17. Silver-bearing veinlet in pyrite (dark grey). It is composed of native silver (white) and freieslebenite (shades of grey). Bar scale = 15 µm.

4.4.1.7 Tellurium

The ore contains trace amounts of several Tellurium (Te) minerals as minute grains in veinlets and as inclusions in pyrite. The minerals are hessite (Ag_2Te), pilsenite (($Bi,Pb)_8(Te,Se)_6$), rucklidgeite (($Bi,Pb)_3Te_4$), and volynskite ($BiAgTe_2$). Since 80.6 % of the Te was recovered in the Cu concentrate, and since all the Te minerals were observed in the Cu concentrate, it is assumed that these minerals were preferentially recovered in the Cu Concentrate.

4.4.1.8. Selenium and antimony

Selenium (Se) is a major element in clausthalite (PbSe) and naumannite (Ag_2Se), a minor element in clausthalite-galena solid solution, and possibly a trace element in chalcopyrite. The low recovery of Se in the Cu concentrate shows that most of the Se occurs in Se in minerals that were rejected to the tails (e.g. clausthalite, galena and naumannite). Hence, only a small proportion of the Se in the ore could occur as a trace element in chalcopyrite.

The antimony (Sb) occurs as a constituent of the minor to trace minerals, freibergite (($Cu,Ag,Fe)_{12}Sb_4S_{13}$), boulangerite ($Pb_5Sb_4S_{11}$) and bournonite ($PbCuSbS_3$), and of the extremely trace minerals, pyrargyrite, dyscrasite, freieslebenite, and costibite. It is assumed that most of the freibergite that was present in the ore was recovered in the Cu concentrate, as the mineral is readily activated by reagents used for Cu flotation, and several grains of freibergite were observed in the Cu concentrate. The low recovery of Sb (18.4%) in the Cu concentrate indicates that most of the Sb occurs as a constituent of boulangerite and bournonite, as these minerals do not float readily in Cu concentrates.

4.4.2. Zn concentrate

Tables 4.4 and 4.5 show a moderate recovery of Zn /sphalerite (82.5 %) in a moderate grade zinc concentrate (47.3 wt % Zn). The Zn concentrate contained 15.3 % of the Au, 21.7 % of the Ag, 45.5 % of the Pb, 3.6 % of the pyrite, 3.5 % of the Cu, 0.9 % of the Te, 7.3 % of the Se, 26.2 % of the Sb, 22.3 % of the Hg and 23.2 % of the Sn.

4.4.2.1. Sphalerite

About 82.5 % of the sphalerite in the ore was recovered in the Zn concentrate. However, since 10.9 % of the sphalerite in the feed was lost in the Cu concentrate, only 89.1 % of the sphalerite in the feed entered the Zn circuit. Therefore, 92.6 % of the sphalerite that entered the Zn circuit was recovered in the Zn concentrate. Of the sphalerite in the Zn rougher concentrate 57 % was apparently liberated, 36 % in particles containing 70 to 99.9 % sphalerite, 4 % in particles containing 30 to 70 % sphalerite, and 3 % in particles with <30 % sphalerite (Table 4.6). The unliberated sphalerite (e.g. 0.1 to 99.9 % sphalerite in particles) was in particles that contained pyrite, pyrrhotite, chalcopyrite, arsenopyrite and gangue. About 14 % of the sphalerite in the Zn concentrate was in particles that contained chalcopyrite inclusions and lamellae (chalcopyrite disease), and 19 % contained pyrite, pyrrhotite and arsenopyrite, and a small amount contained gangue, galena and other minerals. It is noteworthy that most of the galena in the Zn concentrate is present as minute grains attached to sphalerite, which accounts for the higher recovery of Pb in the Zn concentrate than in the Cu concentrate. The Zn concentrate has a low grade because the impurities are present largely as minute to medium sized grains attached to sphalerite, and to a smaller extent, as free or nearly free grains of pyrite and gangue. A few large grains of liberated chalcopyrite were also observed (Figure 4.18). These observations suggest that finer grinding and more cleaning would be required to obtain a higher grade Zn concentrate.

Table 4.6.
Liberation and recovery of sphalerite and galena in Zn circuit

Product	Mineral	Free	70-99.9 % particles	30-70 % particles	<30 % particles
		LIBERATION (%)			
Zn Ro conc	Sphalerite	57	36	4	3
Zn conc	Sphalerite	63	31	4	2
Zn conc	Galena	12	53	11	24
		RECOVERY (%)			
Zn conc	Sphalerite	99	91	62	30
Zn conc	Galena	41	50	39	50

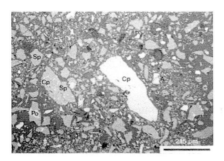

Figure 4.18. Zn concentrate showing a large chalcopyrite grain (white, Cp) as well as small apparently liberated pyrite (white) and gangue grains (very dark grey). Bar scale = 215 μm.

4.5. SUMMARY AND CONCLUSIONS

1. The base metal ores in the Flin Flon and Snow Lake areas in Manitoba are mainly Cu-Zn ores that occur in volcanogenic rocks which have been metamorphosed to the greenschist facies. Metamorphic features of the ore minerals include (1) partial to complete replacement of pyrite grains by pyrrhotite, sphalerite and magnetite, (2) corrosion and rounding of pyrite grains by sphalerite, chalcopyrite and pyrrhotite, (3) reticulate texture of magnetite in pyrite grains, (4) rare dendritic sphalerite and dendritic galena which formed by replacement of pyrite, (5) intensely fractured pyrite, with chalcopyrite and other late minerals filling the fractures, (6) veinlets of remobilised minerals in massive pyrite, (7) porphyroblasts of euhedral pyrite in massive pyrite, (8) augens (eyes) of gangue in massive sulfides, (9) foliated, elongated and schistose chalcopyrite, sphalerite, pyrite and pyrrhotite, (10) diablastic textures caused by intimate intergrowths between chalcopyrite, sphalerite and pyrrhotite, and (11) chalcopyrite with exsolution lamellae of cubanite.

2. The chalcopyrite and sphalerite occur largely as small masses, as relatively large grains, and as interstitial fillings between pyrite grains in massive pyrite. Some of the chalcopyrite and sphalerite are intimately intergrown, and some of the intergrowths are so fine-grained that it would be impossible to separate them by grinding. The sphalerite commonly contains minute blebs and lamellae referred to as "chalcopyrite disease".

3. Most of the gold is present as a Au-Ag-Hg alloy which occurs along pyrite grains boundaries and as veinlets along chalcopyrite-chlorite contacts. Only about 15 % of the gold in the Trout Lake tails occurs as invisible gold in arsenopyrite and pyrite. About 50 % of the gold in the tails is present as exposed gold, and 35 % is encapsulated, probably in pyrite.

4. The Trout Lake deposit contains numerous lenses and each represents a complete hydrothermal cycle. The lenses display the classical gradation of volcanogenic base metal ores, grading from chalcopyrite-pyrrhotite ores at the footwall to sphalerite ores at the hanging wall, and are overprinted by metamorphism. The grade of the Stall Lake ore and its behaviour during mineral processing suggest that the Stall Lake deposit contains the same ore types as the Trout Lake deposit. In contrast the Chisel Lake ore is a sphalerite-galena-pyrite ore type.

5. Since the chalcopyrite in the Trout Lake and Stall Lake deposits is relatively coarse-grained, high Cu recoveries were obtained from these ores by mineral processing (table 4.7). However, the particle sizes of the free and nearly free chalcopyrite in the Cu concentrates were 80 % minus 26 μm, which indicates that the ore had to be ground to a relatively fine size to obtain the high recoveries.

6. The Zn contents in the Cu concentrates obtained from the Trout Lake and Stall Lake ores are relatively high (~4 wt %) and from the Chisel Lake ore is extremely high (13.9 wt %) (Table 4.7). It was noted that (1) the sphalerite in the Cu concentrate has a higher proportion of the so-called "chalcopyrite disease" than the sphalerite in the Zn concentrate, and (2) the unliberated sphalerite in the Cu concentrate was in particles that contained chalcopyrite. It is interpreted that the associated chalcopyrite in the sphalerite-bearing particles was a catalyst for precipitating available Cu ions as a coating on sphalerite surfaces, and thereby activating the particles for flotation in the Cu flotation cells. It is further interpreted that the apparently liberated sphalerite particles that floated in the Cu flotation cells were also enriched in copper, either as chalcopyrite inclusions, not visible on the two dimensional surfaces of polished sections, or in some other form (e.g. Cu in solid solution). The copper in the apparently liberated sphalerite particles would have also been a catalyst for precipitating available Cu ions as a coating on the sphalerite surfaces. It is noted, however, that some of the unliberated sphalerite in the Zn concentrate was in particles that contained chalcopyrite. Therefore, the selectivity of sphalerite particles enriched in copper may have been only partly effective. On the other hand, the Cu cleaners may have rejected many of the selected sphalerite-bearing particles, as the Cu rougher circuit recovered 26.8 % of the Zn in the feed and the Cu cleaner concentrate recovered 10.9 %. The high Zn contents in the Cu concentrates from the ores in the Flin Flon-Snow Lake areas compare with 2.3 wt % Zn in the Cu concentrate from Brunswick Mining and Smelting where the sphalerite is activated by surface coatings of Pb (see Chapter 3).

7. A high Zn recovery (95 %) was obtained from the Chisel Lake ore which is a high grade Zn ore. Moderate (82.5 %) and low (51.9 %) Zn recoveries were obtained from the Trout Lake and Stall Lake ores, respectively (Table 4.7). The sizes of the sphalerite-bearing particles in the Chisel Lake, Trout and Stall Lake Zn concentrates were around 80 % minus 26, 37 and 32 μm, respectively. Hence relatively fine grinding was required achieve the above mentioned Zn recoveries. The grade of the Trout Lake Zn concentrate was only 47.2 %, as it contained free grains of gangue, pyrite and chalcopyrite.

The grades of the feed and concentrates, and recoveries from the Trout Lake, Stall Lake, Chisel lake ores in the Flin Flon-Snow Lake area, and of Brunswick Mining and Smelting ore in New Brunswick are included in Table 4.7 for comparison.

Table 4.7.
Grades and recoveries of metals in concentrates from the Flin Flon-Snow Lake and New Brunswick ores.

	Grade (wt %)				Recovery (%)			
	Trout Lake	Stall Lake	Chisel Lake	BMS	Trout Lake	Stall Lake	Chisel Lake	BMS
Feed								
Cu	2.2	2.4	0.3	0.2				
Zn	2.1	2.5	16.7	9.0				
Pb	0.04	0.2	0.7	4.5				
Au (ppm)	2.2	1.4	2.4	<1				
Ag (ppm)	12.3	14.8	43.9	105				
Cu Conc.								
Cu	28.0	22.5	19.1	22.0	91.7	91.4	42.6	30.8
Zn	4.0	3.9	13.9	2.3	11.0	15.2	0.6	0.1
Pb	0.07	0.18	2.6	15.5	13.1	19.8	2.9	1.5
Au (ppm)	18.7	9.3	53.2		59.2	56.4	16.3	
Ag (ppm)	84.9	112	1047	4740	48.5	77.3	22.0	16.8
Zn Conc.								
Cu	1.7	0.8	0.5	0.2	3.5	0.9	42.9	12.1
Zn	47.3	51.0	54.0	53.6	82.5	51.9	95.0	77.8
Pb	0.4	0.7	0.3	2.2	45.6	20.2	13.3	6.9
Au (ppm)	7.6	1.0	3.8		15.2	1.6	42.7	
Ag (ppm)	59.9	18.2	43.6	78	21.7	3.3	35.0	11.4
Pb Conc.								
Cu			0.3	0.7			1.5	14.2
Zn			10.4	8.9			1.1	7.3
Pb			27.0	34.2			70.5	64.9
Au (ppm)			6.8				5.0	
Ag (ppm)			511	630			24.9	38.4

BMS = Concentrator of Brunswick Division of Noranda, formerly Brunswick Mining and Smelting.

CHAPTER 5

RELATIONSHIPS BETWEEN MINERAL CHARACTERISTICS AND FLOTABILITY

5.1. INTRODUCTION

Information about the influence of mineral characteristics on mineral processing is essential when conducting research on recoveries. The relationship between chalcopyrite liberation and kinetics of chalopyrite flotation was determined by analyzing suites of samples from laboratory tests and from copper cleaner circuits in a commercial concentrator. Similarly, the reasons for low pentlandite recovery from a serpentinized ultramafic nickel ore were determined by studying the relationship between the flotation behaviours of pentlandite and the serpentine content in the ore. A combination of mineralogical and mineral processing techniques was used to determine the behaviuor of pentlandite in the presence of serpentine. The results were integrated, and a sound base was established for selecting reagents that produced a major improvement in pentlandite recovery.

5.2. FLOTABILITY OF CHALCOPYRITE-BEARING PARTICLES

5.2.1. Chalcopyrite in a base metal ore

A base metal ore from Portugal was studied by Professor K.P. Williams, University of Wales, College of Cardiff to determine the flotability of the chalcopyrite in the ore. The ore was ground to minus 150 μm (-100 mesh), and laboratory flotation tests were conducted on the material. Professor Williams submitted three samples for mineral characterization. The samples were:

- Cu rougher concentrate 1 (3.4 wt % Cu). Concentrate 1 was produced in a laboratory rougher flotation cell by processing the ore for 5 minutes,
- Cu rougher concentrate 2 (0.7 wt % Cu). Concentrate 2 was produced by removing all of the concentrate 1 froth, and processing the remaining material in the laboratory rougher flotation cell for an additional 10 minutes (e.g. a total of 15 minutes of processing),
- Cu Tail (0.17 wt % Cu).

The ore minerals in the samples were pyrite, chalcopyrite, sphalerite, arsenopyrite, tetrahedrite, tennantite, bournonite, galena, boulangerite, hematite and magnetite (Lastra and Pinard, 1990).

The Cu recovery in Cu rougher concentrate 1 plus Cu rougher concentrate 2 was less than 70 %. Liberation analyses, performed on the three products (Figure 5.1), show that the particles recovered in concentrate 1 (short flotation time) were apparently liberated chalcopyrite plus particles containing large amounts of chalcopyrite. In particular, 23 % of the chalcopyrite in concentrate 1 was apparently liberated and 79 % was in particles that contained more than 50 % chalcopyrite (including apparently liberated chalcopyrite). The particles recovered in concentrate 2 (longer flotation time) contained less chalcopyrite. Figure 5.1. shows that most of the chalcopyrite in concentrate 2 was in chalcopyrite-bearing particles that contained between 10 and 50 % chalcopyrite, and most of the chalcopyrite in the tails was in chalcopyrite-bearing particles that contained less than 20 % chalcopyrite.

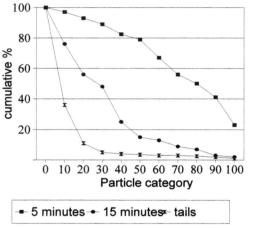

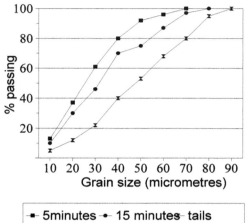

Figure 5.1. Liberations characteristics of chalcopyrite. Plotted as cumulative liberation yield.

Figure 5.2. Size distributions of chalcopyrite grains in chalcopyrite-bearing particles. Plotted as percent passing.

Size analyses showed that concentrate 1 contained the smallest chalcopyrite grains and particles (-70 μm), concentrate 2 contained somewhat larger chalcopyrite grains in chalcopyrite-bearing particles (-80 μm), and the tails contained the largest chalcopyrite grains in chalcopyrite-bearing particles (-90 μm), as well as small chalcopyrite grains in particles with low chalcopyrite contents (Figure 5.2).

5.2.1.1.Conclusions
- Liberated chalcopyrite grains and chalcopyrite-bearing particles containing large amounts of chalcopyrite floated first.
- Unliberated relatively large chalcopyrite grains in chalcopyrite-bearing particles that contain 10 to 50 % Chalcopyrite floated after a long period of processing.
- Large unliberated chalcopyrite grains, and small unliberated chalcopyrite grains in particles with low chalcopyrite contents, did not float.
- The ore that was studied had not been ground fine enough to liberate the chalcopyrite, as most of the unliberated chalcopyrite was coarse-grained.

5.2.2. Test tube flotation test

A suite of samples, prepared from an artificial mixture of chalcopyrite and gangue, was obtained from Dr. Chudacek of M..D. Research Company Pty Limited, North Ryde, NSW, Australia. Dr. Chudacek stated that the samples were prepared using the test tube flotation test (TTFT) procedure with equipment which was designed and manufactured by M.D. Research. The TTFT equipment processes 2 gram samples of ore by simulating the conditions in flotation cells. A first froth is removed after a fixed amount of agitation. The remaining slurry is reconditioned and agitated again, and a second froth is removed. The process is repeated till four froths have been removed. Samples of each froth, and of the ore and residue were studied. The samples were labeled H1, H2, H3, H4, head and residue.

The ore contained 47.9 wt % chalcopyrite, 46.5 wt % quartz, 5.0 wt % pyrite and minor amounts of goethite, sphalerite, cobaltite, amphibole and chlorite. The head sample was a sieved fraction, 25 - 32 μm. The wt % chalcopyrite and weight distribution of each sample are given in Table 5.1. The table shows that the first froth (H1) had the highest grade product and the highest amount of material. The grade and amount of material was lower in each subsequent flotation product (H2, H3 and H4).

Table 5.1.
Chalcopyrite contents and weight percent of samples

	Head	H1	H2	H3	H4	Residue
Chalcopyrite (wt %)	47.9	81.1	32.8	12.5	8.6	5.2
Distribution (wt % of sample)	100	48.5	13.1	8.1	5.0	25.3

Liberation analysis data, presented as cumulative liberation yield curves in Figure 5.3 and as raw data in Table 5.2, show that the chalcopyrite in the feed was largely liberated. Consequently, most of the chalcopyrite in all of the products was also largely liberated (Figure 5.3). However, the proportion of chalcopyrite occurring as apparently liberated grains(100 % particle category) in the residue was considerably lower than in the other products (Table 5.2.).

A detailed analysis of the liberation data (Table 5.2.) shows that the proportion of chalcopyrite occurring as apparently liberated grains (100 % liberation) was highest in the first froth and decreased in each succeeding froth. Furthermore the proportion of chalcopyrite that occurred in particles with less than 90 % chalcopyrite was lowest in the first froth and increased in each succeeding froth.

5.2.2.1. Conclusions
The liberation analyses show that the apparently liberated chalcopyrite floated first, and chalcopyrite-bearing particles floated later, but there was an overlap.

5.2.3. Characteristics of chalcopyrite in cleaner flotation tails of a commercial concentrator

The liberation characteristics of chalcopyrite in the concentrates and tails from cleaner 1 and cleaner 2 flotation circuits in the Trout Lake concentrator were measured (Figure 5.4, 5.5). The results show that 48.5 % and 61.5 % of the chalcopyrite in the Cleaner 1 and cleaner 2 tails, respectively were in particles that contained more than 90 % chalcopyrite (e.g. essentially liberated chalcopyrite). This shows that in order to remove the impurities from the Cu rougher concentrate and from the cleaner 1 concentrate a significant amount of essentially liberated chalcopyrite was rejected. This explains the high assay values commonly observed for the cleaner tails (Table 5.3), and provides a justification for recirculating the cleaner tails (Chapter 4, Figure 4.15).

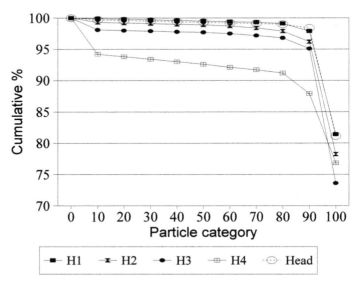

Figure 5.3. Liberation characteristics of samples, plotted as cumulative liberation yield curves.

Table 5.2
Liberation characteristics of chalcopyrite

Particle categories (%)	Head	H1	H2	H3	H4	Residue
100 (apparently liberated)	81.1	81.4	78.2	73.6	76.8	60.9
90 - 99.9	17.3	16.5	18.0	21.5	11.1	33.8
80 - 89.9	0.6	1.2	1.7	1.7	3.3	1.3
10 - 79.9	0.7	0.8	1.4	1.3	3.0	3.2
0.1 - 9.9	0.3	0.1	0.7	1.9	5.8	0.8
Total	100.0	100.0	100.0	100.0	100.0	100.0

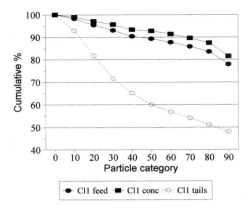

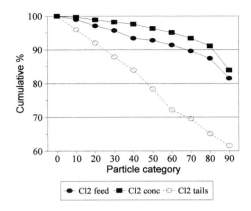

Figure 5.4. Liberation of chalcopyrite in products from the cleaner 1 circuit. Plotted as cumulative liberation yield.

Figure 5.5. Liberation of chalcopyrite in products from the cleaner 2 circuit. Plotted as cumulative liberation yield.

Table 5.3.
Cu assays in mill products, Trout Lake concentrator

Product	Assay (wt % Cu)
Feed	2.16
Cu Rougher Conc.	17.40
Cu Rougher Tail	0.36
Cleaner 1 Conc.	26.29
Cleaner 1 Tail	4.36
Cleaner 2 Conc.	28.00
Cleaner 2 Tail	11.99

5.3. NICKEL IN SERPENTINIZED ORE

The characteristics of pentlandite and of the host rock from the Birchtree deposit in Thompson, Manitoba were studied under a mineral deposits agreement between CANMET (Government of Canada), Manitoba Department of Mines (Government of Manitoba) and the International Nickel Company (INCO). Heida Mani of INCO was the project leader and W. Petruk (CANMET) was the scientific authority. The study was conducted to determine the reason for the low pentlandite recovery from the Birchtree ore. The ore is composed of massive sulfides (pyrrhotite, pentlandite and minor chalcopyrite), seven ultramafic rock types, and metasediments. The ultramafic rocks are composed of serpentine, chlorite, talc, amphibole, mica, carbonates and spinel (mainly magnetite). Serpentine is the main mineral in most of the ultramafic rock types, and forms up to 70 % of the most abundant ultramafic host rock.

Flotation tests, using standard reagents (xanthate), were performed on (1) pure samples of the massive sulfides, (2) barren rocks representing the seven ultramafic rock types and the metasediment, and (3) ores composed of massive sulfides in serpentinized rocks. The results showed:

- a high recovery of pentlandite (92 %) from the massive sulfides,
- negligible flotation of the various rock types from the barren rocks,
- dramatically reduced pentlandite recovery from ores containing massive sulfides in serpentine-bearing rocks.

Subsequent tests, using xanthate, were conducted on artificial mixtures of massive sulfides and rock containing 0 to 45 % serpentine. The results showed that pentlandite recovery decreased with increasing serpentine content in the mixture. The recovery was 92 % from massive sulfides, and decreased to less than 20 % from a mixture containing 45 % serpentine (figure 5.6).

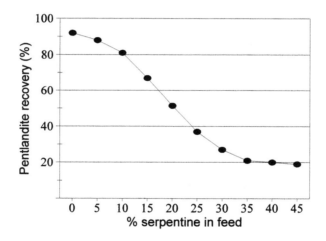

Figure 5.6. Pentlandite recovery as a function of serpentine content in feed (modified from Mani, 1996).

SEM analyses of a pentlandite particle that was lost to flotation tails from an ore with a high serpentine content showed that the surface of the pentlandite particle had numerous submicron grains that contain Mg and Si and are likely serpentine (Figure 5.7). Subsequent zeta potential measurements showed that serpentine and pentlandite are oppositely charged, which indicates that serpentine slime will coat pentlandite. Further flotation tests showed that some reagents such as CMC reversed the charge on serpentine, and the serpentine no longer coated the pentlandite. Hence much higher recoveries of nickel can be obtained from a serpentine-rich nickel ore by using CMC rather than the standard xanthate reagent.

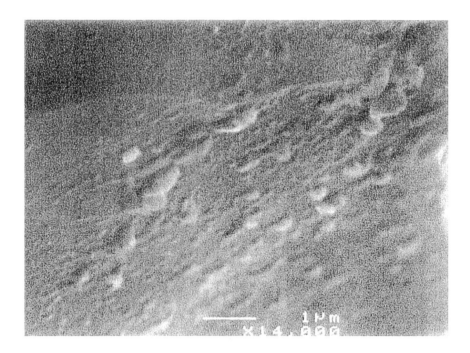

Figure 5.7. Secondary electron SEM photomicrograph of the surface of a pentlandite particle lost to tails. The lighter colored protuberances on the pentlandite surface are minute serpentine grains (slime). Bar scale = 1 μm.

5.3.1. Conclusions

- High pentlandite recoveries were obtained from pentlandite-bearing massive sulfides.
- Pentlandite recovery decreases as the serpentine content in the ore increases.
- Pentlandite recoveries from serpentinized ores decrease because minute serpentine particles (serpentine slimes) coat the pentlandite particles due to opposite charges on the pentlandite and serpentine.
- CMC reagents increase pentlandite recoveries by reversing the charges on the serpentine slimes and reducing the slime coatings on pentlandite.

CHAPTER 6

APPLIED MINERALOGY RELATED TO GOLD

6.1. INTRODUCTION

Applied mineralogy related to gold involves determining mineral characteristics that have a bearing on exploration, mineral processing and hydrometallurgy. Applied mineralogy with respect to exploration involves using ore textures and specific minerals as tracers to gold deposits, and using specific assemblages of silicate and/or ore minerals as indicators of favourable environments for deposition of gold (Hausen et al., 1982).

Applied mineralogy in connection with mineral processing and hydrometallurgy plays a major role when gold is recovered by leaching techniques and flotation. Maximum recovery is a constant goal for gold operators and producers. Knowledge of ore characteristics that affect gold recoveries can help in designing or redesigning a flowsheet, and can indicate whether maximum recoveries have been obtained. Ore characteristics that affect recoveries can be determined by mineralogical techniques, but the techniques need to be designed to analyse gold-bearing materials since gold ores and mill products contain very small quantities of the wanted mineral (e.g. grams of gold per tonne of ore). The mineralogical characteristics that affect gold recoveries are:
- identities of gold-bearing minerals,
- identities of associated metallic and non-metallic minerals,
- proportion of gold occurring in minerals that are soluble in alkaline cyanide solutions (e.g. native gold, electrum, gold alloy),
- proportion of gold occurring as gold tellurides that are soluble in alkaline cyanide solutions at a high pH (>12),
- proportion of gold occurring in minerals that are not soluble in alkaline cyanide solutions (gold selenides and gold sulfides),
- proportion of gold occurring as '*invisible*' gold in other minerals (mainly arsenopyrite and pyrite),
- textures of the gold ores.

This chapter discusses mineralogy of gold, textures of gold in gold ores, the main characteristics of some ore types, methods of characterizing gold ores with respect to processing, methods of processing gold ores, and examples of characterizing tailings from gold operations.

6.2. MINERALOGY

Gold generally occurs as native gold, electrum and gold alloy (Boyle, 1979; Healy and Petruk, 1990d), but some is commonly present as '*invisible*' gold (Cabri et al., 1989), and some as gold tellurides, gold selenides and gold sulfides. In addition the gold in gossan may occur as secondary gold (Petruk, 1990c). Table 6.1 lists the gold-bearing minerals reported by Fleischer and Mandarino (1995). The native gold, electrum and gold alloy form a solid solution series and have similar physical properties and appearances. Therefore, unless qualified, the term gold will be used in this chapter as a generic name which includes native gold, electrum and gold alloy. Gold commonly occurs in association with specific ore and gangue minerals, and some may cause processing problems. The most common associated ore minerals are pyrite, arsenopyrite, galena, sphalerite, chalcopyrite, magnetite, pyrrhotite, and tetrahedrite-tennantite. The common associated gangue minerals are quartz, K - feldspar (adularia, microcline, etc.), albite, calcite, dolomite, ankerite, tourmaline, scheelite, fuchsite, muscovite, chlorite, barite, fluorite, graphite, carbonaceous materials, rhodochrosite, and rhodonite. The main minerals that cause processing problems are graphite, carbonaceous materials, secondary copper minerals (malachite, covellite, chalcocite, azurite, brochantite, etc.), bornite, enargite, marcasite, and pyrrhotite.

Table 6.1.
Gold-bearing minerals

Gold alloys, etc.		Gold Tellurides	
Native gold	Au	Sylvanite	$(Au,Ag)_2Te_4$
Electrum	(Au,Ag)	Kostovite	$CuAuTe_4$
Gold alloy****	(Au,Ag,Hg)	Calaverite	$AuTe_2$
γ- gold amalgam	$(Au,Ag)Hg$	Montbrayite	$(Au,Sb)_2Te_3$
Weishanite	$(Au,Ag)_3Hg_2$	Krennerite	$(Au,Ag)Te_2$
Auricupride	Cu_3Au	Petzite	Ag_3AuTe_2
Tetra-auricupride	$CuAu$	Bilibinskite	$Au_3Cu_2PbTe_2$
Aurostibite	$AuSb_2$	Muthmannite	$(Ag,Au)Te$
Anyuiite	$Au(Pb,Sb)_2$	Bezsmertnovite	$Au_4Cu(Te,Pb)$
Maldonite	Au_2Bi	Bogdanovite	$(Au,Te,Pb)_3(Cu,Fe)$
Zvyagintsevite	$(Pd,Pt,Au)_3(Pb,Sn)$	Buckhornite	$AuPb_2BiTe_2S_3$
Gold sulfides		Gold selenides	
Nagyagite	$Pb_5Au(Sb,Bi)Te_2S_6$	Fischesserite	Ag_3AuSe_2
Uytenbogaardtite	Ag_3AuS_2	Petrovskaite	$AuAg(S,Se)$
Criddleite	$TlAg_2Au_3Sb_{10}S_{10}$	Penzhinite	$(Ag,Cu)_4Au(S,Se)_4$
Buckhornite	$AuPb_2BiTe_2S_3$		
Secondary gold		'invisible' gold (Au as trace component)	
		mineral	gold content (range)
Secondary gold	Au(??)	Arsenopyrite	<0.2 - 15,200 ppm*
Aurantimonate	$AuSbO_3$	Pyrite	<0.2 - 132 ppm*
		Loellingite	<0.2 - 275 ppm**
		Tetrahedrite	<0.2 - 72 ppm*
		Chalcopyrite	<0.2 - 7.7ppm ***

* = Chryssoulis and Cabri, 1990; ** = Neumayr et al., 1993; *** = Cook and Chryssoulis, 1990.
**** = Healy and Petruk, 1990d.

Graphite and carbonaceous materials cause processing problems when the ore is leached to recover gold because they re-absorb some of the gold that has been leached. The phenomenon is known in the industry as preg-robbing (Hausen et al., 1986, Hausen and Bucknam, 1985), as the re-absorbed gold is not available for recovery from the leach liquor.

Minerals, such as pyrrhotite, marcasite, and secondary copper minerals (malachite, covellite, chalcocite, etc.) are soluble in weak alkaline cyanide solutions. These minerals combine with the cyanide and dilute the solution to such an extent that the solution becomes too weak to dissolve the gold. Gold ores that contain significant amounts of gangue minerals, which are soluble in cyanide solutions, cannot be processed by cyanidation without pre-treatment.

6.3. TEXTURES AND MICROSTRUCTURES

Textures that are common to most gold ores are listed below as textures 1 to 3. Textures 4 and 5 have been found mainly in Carlin-type deposits but may occur in other types of deposits. Texture 6 is found only in oxidized zones above orebodies. The main textures for gold in gold deposits are:

(1) gold in fractures and microfractures in rocks, and in veinlets and microveinlets in minerals,

(2) gold in interstitial spaces between mineral grains and at borders of mineral grains, occurring:

    (a) between grains of the same mineral, and

    (b) between grains of two or more different minerals,

(3) gold enclosed in a host mineral (encapsulated gold),

(4) submicroscopic gold associated with framboidal pyrite,

(5) submicroscopic gold associated with very fine-grained clay minerals,

(6) gold in oxidized zones.

6.3.1. Gold in fractures and microfractures in rocks, and in veinlets and microveinlets in minerals

This is the most common occurrence of gold, and the only one where the mineral is visible to the naked eye (Figure 6.1). The gold occurs in mineralized quartz, carbonate, tourmaline and ore mineral veins, which are present as both single veins and stockworks of veins in shear zones and fractures in the rock. The mineralized veins generally contain minor to significant amounts of pyrite, arsenopyrite, chalcopyrite and other sulfide minerals. Some of the gold in the veins may be coarse-grained, and during conventional mineral processing the coarse-grained gold tends to remain on the screens of grinding mills, in pump boxes and in sumps as gold nuggets and gold platelets (Honan and Luinstra , 1996).

Most of the gold in the mineralized veins is, however, present as small grains (Figure 6.2) and micron-thick platelets (Figure 6.3). The small gold grains and platelets occur in veinlets, microveinlets and microfractures within minerals such as quartz, calcite, pyrite, arsenopyrite, chalcopyrite and other sulfide minerals, which are present in the gold-bearing veins. The veinlets and microveinlets are generally narrower than 10 μm, but some may be up to several tens of micrometres wide. Some veinlets are essentially monomineralic and consist solely of gold,

114

Figure 6.1. Hand specimen of very high grade gold ore. The gold (white) occurs as veinlets in quartz. Timmins, Ontario.

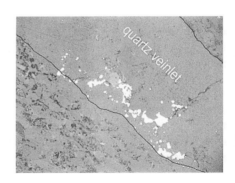

Figure 6.2. Photomicrograph showing a quartz veinlet containing disseminated gold (white). Barberton, South Africa.

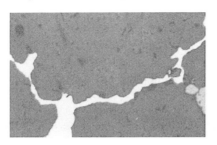

Figure 6.3. Photomicrograph showing a gold veinlet (white) in quartz-carbonate, Doyon Mine, (modified from Guha et al., 1982).

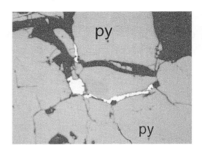

Figure 6.4. Photomicrograph showing a gold veinlet (white) in pyrite (py), Doyon Mine, (modified from Guha et al., 1982).

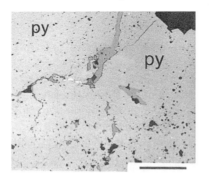

Figure 6.5. Photomicrograph showing a galena veinlet (grey) in pyrite (py). The veinlet contains several grains of gold (white). The pyrite also contains minute inclusions of gold. Trout Lake deposit, Flin Flon, Manitoba. Bar = 160 μm.

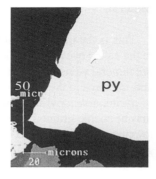

Figure 6.6. Photomicrograph showing encapsulated gold grains (white) in pyrite (py). Hope Brook Deposit, Newfoundland, (from Petruk and Lastra, 1994).

(Figure 6.4), whereas others are composed of several minerals including gold (Figure 6.5). Still other veinlets are composed of assemblages of either late or remobilized minerals including chalcopyrite, tetrahedrite-tennantite, galena, bismuthinite, stibnite, quartz, carbonates, tourmaline and native gold. Some gold also occurs as encapsulated gold in the vein minerals (Figure 6.6), and some may be associated with tellurides, selenides and sulfides (Figure 6.7, 6.8, 6.9).

6.3.2. Gold in interstitial spaces between mineral grains and at borders of mineral grains

Gold commonly occurs at the borders of mineral grains and in interstitial spaces between mineral grains. The most frequent occurrence of border gold is on pyrite grains (Figure 6.10, 6.11, 6.12, 6.13), but it is also common on arsenopyrite (Figure 6.14) and quartz grains, and interstitial gold is present in pyrite, quartz, arsenopyrite, calcite, chalcopyrite and tourmaline (Figure 6.15, 6.16) (Kojonen et al., 1993). Some border gold occurs between two or more grains of the same mineral (Figure 6.11, 6.12, 6.15) and some between grains of different minerals (Gasparrini, 1983). Some border gold is, therefore, present within monomineralic masses, (e.g. in massive pyrite (Figure 6.11), and some is at borders between mineral pairs such as pyrite-quartz (Figure 6.13). The gold grains that occur at mineral boundaries range from 1 to 1000 μm in diameter. Some may be associated with other gold minerals such as gold tellurides, selenides and sulfides (Figure 6.7, 6.14, 6.17).

6.3.3. Encapsulated gold in a host mineral

A small proportion of gold is encapsulated in pyrite, arsenopyrite, loellingite, quartz, chert, and rarely other minerals. The encapsulated gold generally occurs as small, rounded grains that range from a fraction of a micrometer to about 10 μm in diameter (Figure 6.6, 6.19). A few of the encapsulated grains in pyrite, arsenopyrite and rarely in chalcopyrite are part of complex rounded inclusions that are composed of several minerals (Figure 6.18). The complex inclusions are generally smaller than 10 μm, and the gold in these inclusions is around 1 μm. The minerals commonly associated with the gold in such complex encapsulated grains are chalcopyrite, tetrahedrite-tennantite, bismuthinite, galena, rutile, tellurides and other gold minerals. The gold in gold-chert veinlets in some Carlin deposits is often encased in cryptocrystalline chert or in healed microfractures and is not amenable to cyanidation (Hausen, 1985).

6.3.4. Submicroscopic gold associated with framboidal pyrite

Framboidal pyrite occurs in the Carlin deposits as oval to spherical shaped grains that vary from 5 to 20 μm in diameter, and have a microporous texture (Figure 6.20). Aggregates of the pyritic spheroids may locally develop to form a framboidal texture (Hausen, 1981). The spheroidal grains are composed of discrete pyrite crystallites which are clustered to produce a honeycomb texture that has an internal pore space. Gold has not been found in these grains, but microprobe analyses of the mass of spheroidal grains reported 1,000 to 3,400 g/t Au (Wells and Mullens, 1973), and analyses of pyrite crystallites suggest that the pyrite might contain up to 100 g/t '*invisible*' gold (Chryssoulis, 1990). The gold is assumed to be related to the fine-grained texture and high porosity of spheroidal shaped grains (Hausen, 1983).

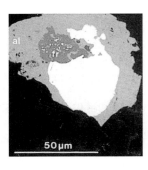

Figure 6.7. Gold (white) assoiated with tellurobismuthite (tb, grey) and pyrrhotite (po, brown) in a quartz-tourmaline vein (black), (modified, from Kojonen et al., 1993).

Figure 6.8. Gold (white) associated with altaite (al) and frohbergite (fr), (modified, from Kojonen et al., 1993).

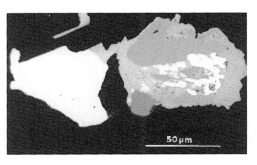

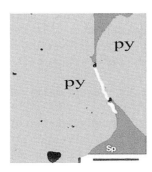

Figure 6.9. Gold (white) associated with tellurides (several shades of grey), Eastern Finland, (modified, from Kojonen et al., 1993).

Figure 6.10. Gold (white) and sphalerite (Sp) at borders of two pyrite (py) grains. Trout Lake deposit, Flin Flon, Manitoba. Bar = 40 μm.

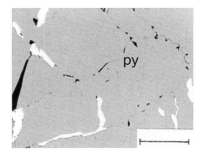

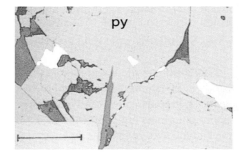

Figure 6.11. Gold (white) in interstitial spaces between pyrite (py) grains. Bar = 50 μm.

Figure 6.12. Gold (white) in interstitial spaces between pyrite (py) grains, and as inclusions in pyrite. Bar = 100 μm.

Figure 6. 13. Gold (white) at edge of pyrite (py) in feldspar porphyry, (modified, from Kojonen et al., 1993).

Figure 6.14. Gold (white) at edge of arsenopyrite (asp) grain in a quartz-tourmaline vein, (modified, from Kojonen et al., 1993).

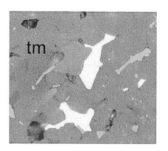

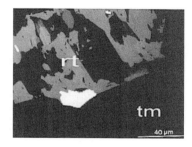

Figure 6.15. Gold (white) and molybdenite (grey) in interstitial spaces between tourmaline (tm) grains, (modified, from Kojonen et al., 1993).

6.16. Gold (white) along rutile (rt)-tourmaline (tm) grain boundary, and as inclusions in rutile, (modified, from Kojonen et al., 1993).

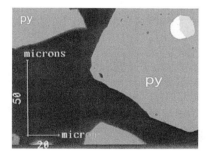

Figure 6.17. Gold (white, Au) at boundary between covellite (cv) and gangue (G).

Figure 6.18. Gold (white) as part of complex gold-chalcopyrite (grey) inclusion in pyrite (py), Hope Brook mine, Newfoundland, (from Petruk and Lastra, 1994).

118

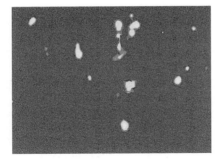

Figure 6.19. Encapsulated gold (white) in quartz (black).

Figure 6.20. Framboidal pyrite from the Carlin gold deposit. Photomicrograph = 65 μm wide, (from Hausen, 1983).

6.3.5. Submicroscopic gold associated with very fine-grained clay minerals

Some gold may occur as very small grains (<0.1 μm) entrapped in particles of very fine-grained clay minerals. Hochella et al. (1987) found such an occurrence in partly oxidized ore from the Carlin mine in Nevada using transmission electron microscopy.

6.3.6. Gold in oxidized zones

The oxidized zones above ore deposits consist of inert primary minerals and of secondary minerals that precipitated during oxidation and decomposition of the ore and silicate minerals. The pyrite, pyrrhotite and arsenopyrite decomposed during oxidation, and the iron from these minerals precipitated as porous secondary iron oxides (e.g. hematite and ferric oxyhydroxides such as goethite and lepidocrocite). It is interpreted that the '*invisible*' gold that was contained in these minerals remained in the oxidized zone as either a gold residue or as secondary gold (see section 3.2.1.6. Gold, Chapter 3). The gold that was initially present as discrete gold minerals in the ore must have remained in the oxidized zone, and presumably was unaltered. The oxidized zones above ore deposits are generally enriched in gold because most of the material that became soluble during oxidation was washed away. Hence the amount of material in the oxidized zone was reduced significantly, but the gold that was in the original ore remained in the oxidized zone.

The gold in some oxidized zones, such as those above porphyry copper deposits, is associated with secondary iron minerals (e.g. Fe oxides, Fe hydroxides, and marcasite) and secondary copper minerals, and some is coated with the secondary minerals. In contrast the most of the gold in oxidized zones above massive sulfide deposits is not associated with secondary copper minerals. The secondary copper minerals above massive sulfide deposits generally occur in a separate zone below the intensely oxidized gold-bearing zone.

6.4. TYPES OF GOLD DEPOSITS

The characteristics of some types of gold deposits are summarized below. The summary does not cover all types of gold deposits, since such coverage is beyond the scope of this chapter. The characteristics of most types of gold deposits have been , however, discussed by Boyle, 1979.

6.4.1 Gold in shear zones

Representatives of gold in shear zones are widespread. The deposits occur as veins, lodes, stockworks, pipes, and irregular masses in extensive fracture and shear zone systems in volcanic, intrusive and interbanded sedimentary rocks of all ages (Boyle, 1979).

Quartz is the most important gangue mineral. Other gangue minerals include calcite, dolomite, ankerite, barite, fluorite, rhodochrosite, rhodonite, adularia, microcline, albite, tourmaline, scheelite, chlorite, sericite and fuchsite. Pyrite is the most common metallic mineral, although arsenopyrite is abundant in many deposits. Other metallic minerals include loellingite, galena, sphalerite, chalcopyrite, pyrrhotite, pentlandite, acanthite, tetrahedrite-tennantite, pyrargyrite, proustite, polybasite, stephanite, miargyrite, stibnite, molybdenite, gold and silver tellurides, other tellurides, gold and silver selenides, native gold, native silver and gold alloys (Boyle, 1979). Oxidized and weathered zones contain marcasite, hematite, rutile, goethite, lepidocrocite, jarosite, and cryptomelane.

Examples of gold deposits in fracture and shear zones are the Kirkland Lake, Porcupine, Red Lake and Hemlo deposits in Ontario; Giant Yellowknife in Northwest Territories; Noranda-Rouyn, Val d'Or, Chibougamau and Belleterre deposits in Quebec; Kolar Goldfields in India; Barberton deposits in South Africa; Mother Lode in Sierra Nevada; Kalgoorlie deposits in Western Australia; Jinqingding gold deposits in China (Xu et al., 1994) and many others (Boyle, 1979).

Much of the gold in these deposits occurs as platelets of native gold along shear zones, fractures and microfractures in quartz and carbonate veins and in the wallrock (Boyle, 1979). A smaller amount occurs as platelets in pyrite, arsenopyrite, and chalcopyrite, and some is present in veinlets in pyrite, arsenopyrite, chalcopyrite and gangue. The veinlets are commonly chalcopyrite, pyrite, calcite, quartz and tourmaline veinlets with small amounts of gold in them (Kojonen et al., 1993). A few are complex veinlets composed of sulphosalts, galena and sphalerite with minute inclusions of native gold. Some gold also occurs along mineral grain boundaries and as interstitial fillings, and some encapsulated gold in arsenopyrite, pyrite, gangue and chalcopyrite is also present. Minor to significant amounts of 'invisible' gold, gold tellurides, gold selenides and gold alloy may be present in these ores.

The gold ores in the Red Lake area in Ontario contain considerable amounts of gold in arsenopyrite as encapsulated gold and as 'invisible' gold. Some of the arsenopyrite crystals are layered and some of the layers are enriched in gold. Some of the layered arsenopyrite crystals contain numerous inclusions of encapsulated gold, whereas others do not.

6.4.2. Gold in Carlin-type deposits

Carlin-type deposits occur in a metamorphosed and faulted sequence of calcareous siltstone, sandstone, silty limestone, chert, and siliceous mudstone (Branham and Arkell, 1995; Livermore,

1996). Some ore horizons are rich in clay minerals such as kaolinite and illite, whereas other horizons are enriched in calcite and dolomite. The ores contain small amounts of pyrite, marcasite, arsenopyrite, and sometimes carbonaceous matter, jarosite and barite. The gold occurs as micron-size grains disseminated within argillized and silicified silty hornfels, marble and siltstone, as small pods of gold along faults, and as fine-grained gold associated with pyrite. Some gold is associated with carbonaceous material, some occurs in interstices of framboidal pyrite, some is encapsulated in chert, a small proportion is present as minute inclusions in large pyrite crystals (Hausen et al., 1986; Hausen, 1981), and sub-microscopic grains of gold (<1 μm) occur in the clay-rich and silicified rocks (Chao et al., 1987; Hochella et al.,1987).

6.4.3. Gold in volcanogenic massive sulfide base metal deposits

Volcanogenic massive sulfide deposits contain from < 1 g/t to ~7 g/t gold. The gold occurs as discrete grains, veinlets, fissure fillings and '*invisible*' gold (Petruk and Wilson, 1993; Healy and Petruk, 1990d, Sinclair, 1990). It is associated with pyrite, arsenopyrite, chalcopyrite, galena, tetrahedrite and sphalerite. The gold associated with pyrite and arsenopyrite occurs as:
(1) minute grains (<1 to ~50 μm) and veinlets along pyrite grain boundaries and in fractures in pyrite and arsenopyrite (Figure 6.10, 6.11, 6.12),
(2) minute grains (<1 to 20 μm) in complex veinlets composed of galena, tetrahedrite and other late minerals, in fractures in pyrite (Figure 6.5),
(3) trapped inclusions in recrystallized pyrite and arsenopyrite,
(4) '*invisible*' gold.
The amount of '*invisible*' gold in pyrite is generally less than 15 ppm, with an average of around 1 ppm (~ 0.7 ppm in the Trout Lake deposit , and 1 to 1.5 ppm in the Mobrun deposit). The average amount of '*invisible*' gold in the arsenopyrite in VMS ores varies from about 20 to 50 ppm.

The gold associated with chalcopyrite occurs as fissure fillings along chalcopyrite-gangue (commonly chlorite and biotite) grain boundaries (Figure 6. 21), and as discrete grains and veinlets in massive chalcopyrite. In some instances the chalcopyrite and sphalerite are fractured and contain gold-bearing veinlets.

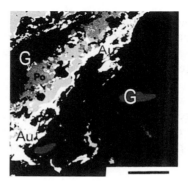

Figure 6.21. Veinlet of gold alloy (Au) (white) in VMS deposit. The gold alloy occurs along the boundary between chalcopyrite (grey), pyrrhotite (po) and gangue (G). Trout Lake Mine, Flin Flon, Manitoba. Bar = 50 μm.

6.4.4. Gold associated with porphyry copper deposits

Many porphyry copper deposits contain gold in the main part of the deposit, and in the nearby surrounding rocks. Porphyry copper deposits consist of stockworks of mineralized quartz veins in intrusive host rocks (commonly granitic) and in associated volcanic rocks. The quartz veins contain inclusions and large grains of chalcopyrite, bornite, pyrite, molybdenite, and magnetite, and trace amounts of a variety of other minerals including galena, sphalerite and gold. The gold is commonly present as micro-veinlets in the quartz, chalcopyrite and pyrite, and possibly as *'invisible'* gold in pyrite and arsenopyrite.

6.4.5. Gold in conglomerate (Witwatersrand-type) deposits

The Witwatersrand deposits in South Africa provide the best examples of gold in conglomerates and reefs. The gold commonly occupies interstitial spaces between the matrix minerals in the conglomerate (Hofmeyer and Potgeiter, 1983), and occurs as irregular, jagged, flaky, plate-like and wire-like grains that are up to 1 cm long and a few micrometers thick. Some of the gold occurs as fracture fillings and as discrete grains in veinlets in fractured pyrite, chalcopyrite, arsenopyrite, cobaltite, uraninite, and quartz. Some also occurs in complex veinlets. The gold in the complex veinlets is often intergrown with authigenic minerals such as chalcopyrite, galena, tennantite, bismuthinite, sphalerite, gersdorffite, and porous pyrite (Oberthür and Frey, 1991; Harley and Charlesworth, 1994). Some of the gold also occurs as interstitial fillings between grains, particulary between pyrite grains (Figure 6.8, 6.9). The least common occurrence is inclusions of gold in large pyrite and quartz crystals. Nevertheless, pyrite is the most important host for the gold, although chalcopyrite is the dominant ore mineral in all reefs.

6.4.6. *'invisible'* gold

Some gold deposits consist essentially of *'invisible'* gold in pyrite and arsenopyrite and are, therefore, classified as refractory gold deposits. The Suurikuusikko Au deposit in central Lapland, which was studied by Kojonen and Johanson (1999) is an example of such a deposit. The average *'invisible'* gold contents in pyrite and arsenopyrite are about 46 ppm and 279 ppm, respectively. The distribution of gold in the ore is 4.1 % free gold, 22.7 % *'invisible'* gold in pyrite and 73.2 % *'invisible'* gold in arsenopyrite. The gold content in pyrite varies from <22 to 585 ppm, and in arsenopyrite from <22 to 964 ppm. The pyrite grains are zoned and there is a general correlation between the Au and As contents in pyrite. In contrast there is a negative correlation between the Au and Sb contents in arsenopyrite (Kojonen and Johanson, 1999) .

Another deposit that contains high amounts of *'invisible'* gold in pyrite and arsenopyrite is a section of the Olympias deposit in Greece. Samples of the ore were analyzed by Chryssoulis and Cabri (Cabri and Chryssoulis, 1990; Chryssoulis and Cabri, 1990). The average *'invisible'* gold content in arsenopyrite is 49 ppm, in arsenian pyrite is 49.6 ppm, and in arsenic-poor pyrite 3.4 ppm. The distribution of gold in the ore is 9 % free gold, 43.8 % *'invisible'* gold in arsenian pyrite, 15.7 % *'invisible'* gold in As-poor pyrite, and 31.5 % *'invisible'* gold in arsenopyrite. There is a general correlation between the As and Au contents in pyrite, with the highest *'invisible'* gold content in pyrite being 132 ppm.

Fleet et al. (1993) have shown that '*invisible*' gold in arsenian pyrite is associated with As-rich growth bands, and there is a general correlation between the As and Au contents in arsenian pyrite, although some As-rich bands do not contain Au. The highest Au content that they found in arsenian pyrite was 1400 ppm in pyrite from an arsenic-rich band in pyrite from the Fairview mine near Barberton in East Transvaal, South Africa.

6.5. CHARACTERIZING GOLD ORE WITH RESPECT TO PROCESSING

Gold ores are characterized with respect to processing to determine the maximum possible recovery, to determine the reason for high losses of gold to tailings, and to predict how the gold minerals would behave during processing. One technique of characterizing a sample of gold ore was described by Chryssoullis and Cabri (1990) and modified by Petruk et al. (1995), another was described by Tassinari (1996), and a third by Kojonen and Johanson (1999). All techniques have the common approach of identifying the minerals and, where possible, determining mineral quantities. The first technique also determines the amount of leachable gold by grinding the material to about minus 10 μm, and performing cyanidation tests on the ground material. The technique described by Tassinari (1996) determines the amount of recoverable gold by using amalgamation and cyanidation tests, and the technique described by Kojonen and Johanson (1999) determines the distribution of gold among the minerals by using diagnostic leaching. In addition to the above techniques, the amount of '*invisible*' gold in pyrite and arsenopyrite may need to be determined. A sequence of steps, which utilizes parts from all techniques, is proposed for a full characterization of gold ores with respect to processing. In practice some of the steps would not be performed as they may duplicate other steps. The proposed steps are:

- select a sample, sample fractions and sub-fractions for analysis,
- sieve sample,
- assay the sample, the fractions and the sub-fractions,
- identify the minerals,
- determine mineral quantities,
- determine proportion of exposed and encapsulated gold in each sieved fraction,
- grind sample and leach to determine maximum amount of leachable gold,
- perform diagnostic leaching and, where possible, amalgamation and cyanidation tests,
- analyze for '*invisible*' gold,
- calculate a mineral balance for gold among different minerals,
- interpret the results.

6.5.1. Selecting samples and fractions for analysis

The sample studied may be a tailings sample or a head sample which had been crushed to the size of feed to the grinding circuit. About 4 kg of material is required. The material is split into four 1 kg sub-samples. One sub-sample is used for a mineralogical study, the other three sub-samples are used for leaching tests. The sub-sample selected for mineralogical analysis is sieved, and a nominal sized fraction is selected for detailed study. The selected fraction should have particles that are large enough to handle under a binocular microscope, and small enough to contain many grains of heavy minerals. A common sieve size is 65 to 100 mesh (212 to 150 μm). In some instances when the material has been finely ground a 270 to 325 mesh (53 to 44 μm) fraction has to be used.

The selected sieved fraction is separated, with a heavy liquid that has a specific gravity of 3.33, into float and sink sub-fractions to concentrate the gold minerals. This step is essential because gold is generally present in very small amounts in the samples studied. The heavy liquid sink fraction should contain enough grains to represent all the gold species in the sample.

6.5.2. Assaying the sample and fractions

The unseparated head sample, the sieved fractions, and the sieved heavy liquid float sub-fraction are assayed for gold. The gold content in the heavy liquid sink sub-fraction is calculated by difference, as often there is not enough sink sub-fraction for assay. The reason for analyzing both the unseparated head sample and the sieved fraction selected for study is to note whether there is a wide difference in gold content between them. If the difference is small it is assumed that the sieved fraction is more or less representative of the sample. The assay of the heavy liquid float fraction provides an indication of the proportion of gold associated with or encapsulated in silicate minerals at the specific sieve size.

6.5.3. Identifying the minerals

The minerals that need to be identified are gangue minerals, sulfide minerals, gold minerals, potential gold carriers such as pyrite and arsenopyrite, and minerals that interfere with cyanidation (e.g. carbonaceous material, graphite, marcasite, pyrrhotite, secondary copper minerals and clay minerals). The gold minerals are native gold, electrum, other gold-alloys, Au tellurides, Au selenides, Au sulfides, uncharacterized secondary gold minerals or phases which could be Au hydroxides or similar phases that precipitate from solutions at low temperatures. Special attention is paid to detect the presence of carbonaceous materials, graphite and clay minerals. As noted above (section 6.2) the carbonaceous materials and graphite can absorb gold that has been dissolved by cyanidation, and hence rob it from the pregnant solution (Hausen and Bucknam, 1985). Clay minerals can contain inclusions of colloidal-sized gold grains and shield them from the leaching liquor (Hochella et al., 1987).

The major silicate minerals are usually identified by analyzing the crushed powders (heavy liquid float sub-fraction) by x-ray diffraction. Most of the minor and trace minerals are identified by analyzing polished sections of the heavy liquid sink sub-fraction with an ore microscope and a scanning electron microscope equipped with an energy dispersive x-ray analyzer. Unusual minerals, including rare gold minerals, are identified with an electron microprobe. In addition selected grains may be identified by X-ray diffraction analysis, transmission electron microsocopy, or other techniques.

6.5.4. Determining mineral contents

Mineral quantities and relative proportions of the minerals in the heavy liquid sink sub-fractions are determined by image analysis or point counting, and in heavy liquid float sub-fractions by X-ray diffraction analysis. In special cases, a combination of image analysis (or point counting), X-ray diffraction analysis and calculations from chemical assays is used. Special precautions are taken to accurately determine the quantities of minerals that might contain 'invisible' gold (pyrite and arsenopyrite).

If only a few gold grains are found by optical and scanning electron microscopy, a gold search is performed with an image analyser to find enough gold grains to determine the distribution of gold among the different gold minerals.

6.5.5. Proportion of exposed and encapsulated gold in each sieved fraction

The proportion of gold that occurs as apparently liberated, exposed and encapsulated grains in each sieved fraction is determined to predict the proportion of recoverable gold. The analysis is performed by making a polished section of each sieved fraction, and analysing each polished section by image analysis or point counting. Image analysis would be performed by using a search technique to find the gold grains, and using a macro that will measure the size of each gold grain and determine whether it is apparently liberated, exposed or encapsulated.

Another approach is to use amalgamation and cyanidation tests (Tassinari, 1996). This approach is seldom used as amalgamation is not permitted in many countries. The amalgamation-cyanidation approach involves:
(1) conducting diagnostic amalgamation tests on sized gravity fractions to determine the amount of apparently liberated and nearly liberated gold,
(2) cyanidation tests on the amalgamation tailings to determine the amount of exposed gold,
(3) assaying the cyanidation tailings to determine the amount of encapsulated gold.

6.5.6. Amount of cyanidable gold

The maximum amount of cyanidable gold is determined by re-grinding the remaining three 1 kg sub-samples of the ground feed sample for about 5 to 6 hours in laboratory ball mill to reduce the material to about 95% minus 10 to 15 μm. One batch is cyanided to determine cyanide consumption. A second batch is cyanided at a moderate pH (10.5 to 11) to determine the amount of cyanidable gold (i.e. native gold, electrum, gold-alloys and metastable gold), and a third batch at a high pH (12 to 12.5) to dissolve the native gold, electrum, gold alloy and soluble gold tellurides (Chryssoulis and Cabri, 1990).

At the intense regrind (5 to 6 hours) most of the gold minerals would be exposed to cyanide solutions because most gold mineral grains, as small as 1 μm, tend to occur along fractures and grain boundaries which are the loci for rock and mineral breakage (Petruk, 1995). Only grains that were trapped as inclusions in other minerals during recrystallization or primary crystal growth, might remain encapsulated. The proportion of gold minerals that remain encapsulated at this fine regrind is not known, but observations at CANMET suggest that for most ores it is low.

The proportion of gold that is cyanided at a moderate pH usually accounts for most of the gold that is present as soluble gold minerals (i.e. native gold, electrum, gold alloys and metastable gold), and the proportion of gold cyanided at high pH gives an indication of the proportion of gold occurring as tellurides. The remainder of the gold is assumed to be present as Au selenides, Au metalloids, Au sulfides and '*invisible*' gold.

6.5.7. Diagnostic leaching

Steps 1 to 4 of the following diagnostic leaching test, wereas described by Lorenzen and Tumilty (1992) and subsequently used by Kojonen and Johanson, (1999). The author has frequently used step 5.

Step 1. Cyanidation to determine the amount of free gold.

Step 2. Leach with hydrochloric acid to define gold bound to carbonates, pyrrhotite, galena, goethite.

Step 3. Leach with sulphuric acid to determine amount of gold associated with uraninite, sphalerite, labile copper sulphates, labile base metal sulfides, and labile pyrite.

Step 4. Leach with nitric acid to determine amount of gold associated with pyrite, arsenopyrite and marcasite.

Step 5. Leach with HF to determine amount of gold encapsulated in silicates.

6.5.8. *'Invisible'* gold (microbeam analysis and staining)

Gold that could not be observed with a an optical microscope and scanning electron microscope, but is present as a trace element in some minerals, particularly arsenopyrite and pyrite, is referred to as *'invisible'* gold. The concentration of *'invisible'* gold in the minerals is determined by analyzing the minerals (pyrite and arsenopyrite) by Secondary Ion Mass Spectrometry (SIMS) or a low detection microprobe technique. SIMS is the most sensitive microbeam technique for measuring trace element contents. The detection limit for *'invisible'* gold in pyrite and arsenopyrite is about 0.2 to 0.3 ppm by SIMS, about 20 ppm by the low detection microprobe technique, and about 200 ppm by standard microprobe techniques. Arsenopyrite is the most common and probably the most significant carrier of *'invisible'* gold, with analyzed concentrations ranging from less than 1 to 15,200 ppm (Chryssoulis and Cabri, 1990) (commonly 20 to 50 ppm). Pyrite is another common carrier of *'invisible'* gold, but concentrations are much lower (<0.3 to 1400 ppm (Fleet et al., 1993)), commonly 0.1 to 5 ppm, as in the Mobrun deposit, NW Quebec (Larocque et al., 1995). In some deposits, loellingite ($FeAs_2$) is a significant gold carrier (Neumayr et al., 1993). All other metallic minerals have low contents of *'invisible'* gold, although tetrahedrite has been found to contain up to 59 ppm (Cook and Chryssoulis, 1990).

Generally 20 to 50 grains of a mineral are analyzed to determine the average amount of *'invisible'* gold. The quantities of pyrite and arsenopyrite in the sample are determined by image analysis, or by point counting coupled with assays for Fe, As and S.

The distribution of *'invisible'* gold in the arsenopyrite, pyrite, and loellingite grains can be observed by SIMS imaging (Fleet et al., 1993). SIMS images may provide guidelines to processing and/or metallurgy by showing whether the *'invisible'* gold is homgeneously distributed throughout the mineral or is concentrated in zones. If the gold-bearing zones occur along grain boundaries it may be possible to separate the grains of the gold-bearing mineral (arsenopyrite) from barren grains of the same mineral. Fleet et al. (1993) have also shown that the colour of a potassium permanganate stain on pyrite in polished sections can be correlated with the distribution of As in pyrite and indirectly with *'invisible'* gold, since there is a general correlation between the As and Au contents in some pyrite grains. Kojonen and Johanson (1999) have successfully applied the technique. The technique that they used was to dissolve $KMnO_4$ in concentrated sulphuric acid (Schneiderhöhn, 1952; Ramdohr, 1975), and etch the mineral for 1 to 2 minutes. The sections were rinsed in flowing water and dried with a hair drier. Immediately after drying the sections were studied under reflected light and photographed.

6.5.9. Calculating mineral balance for gold among different minerals

The distribution of gold among the different minerals is calculated as follows:
- the amount of gold that was analysed as soluble gold at a moderate pH by the grinding and cyanidation test is equal to the maximum amount of cyanidable gold,
- the amount of gold analysed as soluble gold at high pH minus the amount of soluble gold at moderate pH is equal to the amount in gold tellurides,
- the amount of free, exposed and encapsulated gold in each size fraction is calculated from ratios determined by image analysis for free, exposed and encapsulated minerals,
- the amount of encapsulated gold in each mineral can also be determined by the diagnostic leaching tests,
- the amount of recoverable gold can be determined by amalgamation and cyanidation tests,
- the amount of '*invisible*' gold occurring in each mineral is calculated from the average amount of gold in each mineral, as determined by SIMS analysis or low detection microprobe analysis, and from the mineral quantities determined by image analysis,
- the remainder is unaccounted, and could represent an analytical error or unidentified gold carriers such as Au selenides, Au sulfides and other insoluble gold minerals.

6.6. PROCESSING GOLD ORES

Gold ores are processed by alkaline cyanide leaching, heap leaching, flotation and gravitational techniques.

6.6.1. Leaching (Cyanidation)

Cyanidation is the most common technique for recovering gold. It is performed by leaching the ore in an alkaline solution that has a low concentration of alkaline cyanide (cyanidation). Other solutions such as thiourea, nitric acid, halides, etc., have been tested and used in special cases, but alkaline cyanide (Na-cyanide) is generally used because it is chemically robust and usually is very forgiving of non-optimum operating conditions. Furthermore, despite its toxicity, the cyanide ion is easy to oxidize and rendered harmless in gold plant tailings.

The alkaline cyanide solution dissolves gold and silver at a pH of 10 to 11, and dissolves some gold tellurides at a higher pH (~12). Cyanidation is, however, usually performed at a pH of 10 to 11 hence, only native gold, electrum, gold alloy, secondary gold and native silver are dissolved.

Until recently gold was precipitated from the leach liquor by zinc cementation. In the last twenty years carbon adsorption processes, which include carbon in pulp (CIP), carbon in leach (CIL) and carbon in columns (CIC), have replaced many zinc cementation plants (Fleming,

1998). The First commercial use of the CIP process was at the Homestake Mine in South Dakota in 1973 (Hall, 1974). It was found to be considerably more economical to install and operate than the zinc cementation process, had a higher gold recovery, and was less vulnerable to impurities such as sulfides, arsenates and antimony in the leach liquor.

The CIP process recovers gold directly from a pulp or slurry that contains 50 to 60 % solids. The gold is leached in one tank, and is adsorbed onto activated carbon in another tank. The loaded carbon is recovered by a screening device which has a screen mesh size that allows the gold depleted pulp to pass through while retaining the carbon granules. The CIL process operates in a similar manner to the CIP process but the activated carbon is added to the leaching tanks, and adsorption occurs simultaneously with the leaching. The CIC process operates by pumping the pregnant leach liquor upflow through a series of columns that are packed with activated carbon (Flemming, 1998).

6.6.2. Refractory gold ores

When the ore does not respond well to direct cyanidation it is prefrred to as a refractory ore. The actual leach efficiency is somewhat arbitrary, although most persons in the gold industry consider that an ore is refractory when less than 80 % of the gold is recovered by cyanidation (Flemming, 1998). An ore may be refractory because:
- the gold occurs in minerals that are not soluble in cyanide solutions,
- the gold occurs as 'invisible' gold',
- the gold is encapsulated as minute inclusions in other minerals, and the cyanide solutions cannot come into contact with the gold inclusions,
- the ore contains graphite and or carbonaceous material that cause preg-robbing,
- the ore contains minerals that use up the cyanide solution and dilute it to such an extent that the solution is too weak to dissolve all the gold,
- impurities may coat surfaces of gold particles and prevent the gold from being leached,
- gold nuggets may be too large to dissolve within the prescribed residence time.

Refractory gold ores are pre-treated by roasting, pressure leaching or bioleaching prior to cyanidation.

6.6.2.1. Roasting

Roasting has been used for many years to recover the gold from refractory gold ores and is still widely used. When arsenopyrite is present in the ore a two-stage process is usually applied. A non-oxidizing first stage roast at 400-450^{0}C is performed to remove the arsenic as volatile arsenic trioxide, followed by an oxidizing roast at 650 to 750^{0}C, to produce a permeable hematite and SO_2 (Fleming, 1998). The 'invisible' gold and insoluble gold minerals are converted to gold that is soluble in a cyanide solution, and the native gold and electrum remain in the residue as soluble gold minerals. The gold that was encapsulated in the sulfides and arsenopyrite is now in the permeable hematite and can come in contact with the cyanide solutions and be dissolved. The roasting may burn off the graphite and carbonaceous material and eliminate the preg-robbing characteristics of the ore. In some instances, however, residual carbon might be left and sometimes it may be a more active variety than the original one in the ore. A major drawback of roasting is that it is difficult to condense the volatile arsenic trioxide and to filter it from the off-gas.

The gold ores from the Giant Yellowknife mine, Northwest Territories, Canada, from the Red Lake deposits in Ontario, Canada, and from the Kalgoorlie deposits in West Australia, provide examples of ores that require roasting. The gold in the Giant Yellowknife mine occurs largely as submicroscopic gold associated with arsenopyrite. Only about 30 % is recoverable by cyanidation. Roasting of arsenopyrite and cyanidation of the roasted calcine is required to recover the remaining gold (Thomas et al., 1987). Similarly, some of the gold in the Red Lake deposits in Ontario is associated with arsenopyrite. Gold-bearing arsenopyrite concentrates are produced to recover the gold associated with arsenopyrite. The arsenopyrite concentrates are roasted at a commercial smelter and the roasted calcine is cyanided.

6.6.2.2. Pressure leaching
There has been a shift in recent years from the traditional roasting method of treating refractory gold ores to pressure leaching. A higher gold recovery is obtained by cyaniding the pressure leach residue than the roasted calcine, and the process produces a very stable ferric arsenate complex ($FeAsO_4$) (Fleming, 1998). Both pyrite and arsernopyrite are decomposed by pressure leaching and occluded gold is exposed. The '*invisible*' gold is released from the arsernopyrite and pyrite and it precipitates as secondary gold which is soluble in the cyanide solutions. Cyanidation of the residue will recover the gold that has been exposed, as well as the gold the has been released from the mineral structure. Some of the iron and sulfur that were released by the decomposition of the sulfides precipitate as hematite, ferric sulfate and jarosite.

6.6.2.3. Bioleaching
Bacterial leaching is a relatively new technique of processing refractory gold ores, and is being investigated by many companies. The technique has the same advantages as pressure leaching in that the pyrite, arsenopyrite and other sulfides are decomposed and the occluded gold is exposed. Similarly the '*invisible*' gold is released and precipitates as secondary gold that is soluble in the cyanide solution. The iron, arsenic and sulfur precipitate as relatively stable compounds. A major concern of bioleaching has been the residence time. Early pilot plant work at the Gencor Laboratory in South Africa required a 10 day residence time to achieve sufficient oxidation of an arsenopyrite concentrate for a 97 % gold recovery by cyanidation of the bioleach residue. After two years of operation the bacteria had adapted and mutated to the extent that the retention time had decreased to 4 days (van Aswegen et al., 1988). In contrast, a typical pressure leaching operation requires a residence time of 1 to 2 hours. It is noteworthy that processing tests on the refractory gold ore from the large Suurikuusikko gold deposit in Finland gave recoveries of 10 % without bioleaching and 96 % with bioleaching (Härkönen et al., 1999).

6.6.3. Flotation

Gold is generally recovered by flotation when it is a by-product in sulfide ores, as in porphyry copper, base metal and copper-gold ores. The gold is generally recovered in the copper concentrate, and is recovered from the copper concentrate by smelting and electrolysis. Gold tends to float readily in sulfide flotation cells, particularly in copper flotation cells. Hence all the liberated gold, and much of the unliberated but exposed gold is recovered in copper concentrates. Much of the attached gold is recovered, particularly in the rougher concentrate, even if it is attached to other minerals, such as pyrite. Unfortunately some of the attached gold that is recovered in the rougher cells may subsequently be lost in the cleaning stages, because the attached gold grains may be too small to maintain flotation of the particles. However, a significant amount of the gold commonly remains in the copper concentrate, hence much of the gold lost to flotation tailings is encapsulated gold grains.

At some operations, as at the Kutema gold mine in Southern Finland, gold is recovered by flotation to produce a sulfide concentrate that can be smelted to recover the gold. The Kutema gold ore consists of disseminated, banded to massive pyrite with various tellurides and minor base metal sulfides, arsenides and sulphosalts in a sericite-quartz schist. The gold occurs as inclusions in the pyrite, arsenopyrite and quartz, intergrown with tellurides and as free grains. About 82 to 87 % of the gold is recovered (Kojonen et al., 1999).

In some cases flotation is used as a scavenger to recover the gold that was not recovered by cyanidation because the gold was encapsulated in the sulfides (pyrite an chalcopyrite) and did not dissolve during cyanidation. In one case, which cannot be identified because of company confidentiality, the gold occurred in quartz veins and was associated with pyrite and chalcopyrite. The ore was crushed, ground and cyanided. About 85 % of the gold was recovered by cyanidation. A flotation circuit was installed to process the cyanidation tailings to recover more gold. The cyanidation tailings were reground and a copper concentrate, grading around 22 wt % Cu, recovered around 30 % of the gold in the tailings. A study of the products in the flotation circuit showed that some relatively large gold grains, attached to pyrite, were lost to the tailings. The flotation practice was changed to recover particles containing exposed gold attached to pyrite. The grade of the copper dropped to 17 wt % Cu, but the gold recovery increased by 8 %. The smelter accepted the copper concentrate.

In other cases flotation is used prior to cyanidation to either pre-concentrate the ore, or to remove minerals such as secondary copper minerals from the cyanidation circuit. Flotation is used to pre-concentrate the ore only when there is a high recovery of gold in the pre-concentrate. If the pre-concentrate does not recover most of the gold, all the ore is processed by cyanidation.

6.6.4. Gravitational techniques

Gravity concentration techniques are used to supplement cyanidation and flotation techniques for recovering gold (Laplante et al., 1996), particularly when the ore contains gold nuggets that are too large to attach to bubbles in flotation cells, and too large to be dissolved completely during cyanidation. Evidence of gold nuggets in an ore is usually found by their presence on the crusher plates, grinding mill liners, grinding pump boxes and sumps. The gravity circuit in most gold concentrators is placed ahead of the cyanidation circuit, and a high grade gold concentrate is usually recovered. The Golden Giant Mine of Hemlo Mines Inc. provides an example of a gravity circuit ahead of the leaching circuit. The gravity circuit consists of Knelson concentrator and shaking tables (Honan and Luinstra, 1996), and recovers a gold concentrate that contains about 75 wt % Au.

6.6.5. Heap Leaching

Heap leaching of gold ores became a widely used technology in the gold mining industry in the 1990's. It is a low-cost method of recovering gold from low grade materials with recoveries of about 50 to 90 % of the contained gold. The technique involves percolating an alkaline cyanide solution through a heap and collecting the pregnant solution. The basic requirement is that the heap be porous enough for the cyanide solution to flow through, and that the ore pieces be permeable so that the solution can come in contact with the gold. This criteria is met by material from oxidized ore zones above ore bodies, and by some primary ores. Bioleaching is used increasingly to release the gold contained in sulfides, and in some instance fine-grained ores are cast into briquettes to make them permeable. Investigations have shown that heap leaching can be performed in winter in permafrost conditions (Lakshmanan and McCool, 1987), and year round heap leaching operations have been developed in areas of severe winter conditions (Micheletti and Weitz, 1997; Komadina and Beebe, 1997; Smith, 1997).

6.6.5.1. Vein type deposits
It is inferred that most of the gold in the vein type deposits is amenable to cyanidation and heap leaching, and high recoveries might be obtained by heap leaching, particularly in material from oxidized and weathered zones above the deposits.

The operations at the McDonald gold deposit in Nevada provide an example where processing an ore from a vein type deposit is done by heap leaching (Enders et al., 1995). Gold and silver mineralization occurs in veins in volcanic rocks, mainly in tuffs, and some is disseminated in the wall rock. The deposit has been oxidized to depths of more than 300 m (985 ft). About 186 M tons grading 0.9 g/t was proposed to be mined by open pit and processed by heap leaching. Metallurgical tests, which included bottle roll tests, small diameter column tests, and pilot plant tests, indicated that the ore can be processed economically by heap leaching. The bottle roll tests indicated that lithology, oxidation intensity, and vein density had the greatest impact on gold recovery. Gold recovery ranged from 68 % on -150mm (-6 inch) material to 90 % on -6mm (-0.2 inch) material. These properties suggest that most of the gold occurs as veinlets in fractures, and a significant proportion is associated with pyrite. When the pyrite was oxidized, the contained gold became amenable to leaching.

Another example of heap leaching a vein type gold ore is the operation at the Cresson mine in Cripple Creek, Colorado (Jeffrey and Joseph, 1996). Most of the gold occurs in veins and veinlets and as disseminations in volcanic rocks, but some is present as gold tellurides and some is encapsulated in pyrite and probably quartz. The deposit has proven and probable reserves of 62.7 M tons grading 0.93 g/t. Average heap leach recovery was established at 70 %.

Heap leaching tests on fresh ore from Larder Lake, Ontario have shown a 45 % recovery on 3/8 inch material, and 70 % recovery on 1/4 inch material (Witte and Witte, 1985). The high recovery on 1/4 inch material indicates that much of the gold occurs in veinlets, and the 1/4 inch material has a relatively high permeability.

6.6.5.2. Carlin-type deposits
Oxidized zones of Carlin-type deposits in Nevada (Rota, 1997) and of the Black pine gold deposit in Idaho (Clemson, 1988) are amenable to heap leaching yielding 50 to 70 % recovery of contained gold (Rota, 1977). The gold in the oxidized zones is amenable to leaching because the pyrite has been converted to porous hematite, and the gold that was associated with pyrite is available for dissolution by the cyanide solution. On the other hand, it is likely that the textural

relationships between gold and silicates (particularly quartz) did not change during oxidation. Chryssoulis and Wan (1977) studied the tailings from a leaching operation and found minute inclusions of gold encapsulated in silicates.

6.6.5.3. Volcanogenic massive sulfide base metal deposits
Oxidation and weathering of some volcanogenic massive sulfide orebodies has altered the sulfides in the upper parts of orebodies to a gossan that is composed largely of goethite and lepidocrocite, and to a copper-rich zone below the gossan. The gossan is enriched in gold, hence the gossans in the Bathurst area in New Brunswick and at the Rio Tinto deposits in Spain contain ~2 g/t gold, and some have been mined for the gold. The gossans are amenable to heap leaching.

6.6.5.4. Porphyry copper deposits
The Mount Poly (Chong et al., 1991) and Mt. Milligan deposits (Mellis et al., 1991) in British Columbia, Canada, are two porphyry copper-gold deposits that were tested for heap leaching of gold from oxidized ores above the deposits. The Mount poly deposit contains 52 M tons of ore grading 0.38 wt% Cu and 0.55 g/t Au. The Mt. Milligan deposit contains about 449 M tons of ore grading .19% Cu and 0.44 g/t Ag. Both orebodies have a supergene zone that contains malachite, chrysocolla, native copper, cuprite, digenite and covellite. Cyanidation tests indicate that the gold is amenable to cyanidation, but the cyanide consumption is very high due to the presence of soluble copper minerals. Tests have subsequently been conducted to recover both the Cu and Au by conventional flotation.

The Andacollo Gold deposit in Chile is peripheral to the Andacollo porphyry copper system, and occurs in Lower Cretaceous volcanic rocks (Bernard, 1996). The mineralization is in vertical structural breccia zones or hydrothermal feeders. The mineralized breccia is characterized by potassic alteration and contains pyrite and magnetite. Gold is associated with pyrite but is not incorporated in the pyrite structure, hence it leaches freely in cyanide solutions. Laboratory column leach tests indicate that gold recovery increases with decreasing particle size, and the optimum economic heap leach feed size is -9.5 mm (-1/4 inch). Recoveries ranged from 74 % to 82 % in tests conducted on three heap sites containing 7,500 short tonnes of -9.5 mm material. As expected the gold recoveries did not correlate with ore grade, rock type or deposit geometry, since these factors do not relate to ore textures which influence recoveries by heap leaching. Minable ore reserves in 1966 were 36.5 M tons with an average grade of 1 g/t. Reagent consumption was moderate.

The Kennecott Barneys Canyon Mining Company is mining and processing a porphyry copper deposit that is peripheral to the Bingham copper deposit (Braun and LeHoux, 1993). The operation (LeHoux, 1997) involves floating the sulfide minerals and agglomerating the tailings. The agglomerated tailings are processed by heap leaching to recover the gold.

6.6.5.5. Bioleaching
Bioleaching of a heap of low grade sulfidic refractory material prior to heap leaching presents a low cost method of recovering the gold from refractory material. Newmont gold company has developed, demonstrated and patented a biooxidation pretreatment process (Shutey-McCann et al., 1997). The process allows the company to heap leach sulfide-bearing refractory gold ores containing 1 g/t to over 3 g/t Au. The process involves developing and growing culture bacteria

to be used as inoculum on the heaps. The demonstration of the process involved crushing the ore to minus 19 mm, agglomerating it with inoculum, and stacking it into heaps. Subsequently an average of 18 litres of inoculum per ton of ore was added, and the temperature of the biooxidation activity was controlled. After a certain period the biooxidation was stopped, and cyanide heap leaching was performed to recover the gold. Other approaches to heap bioleaching have been proposed (Bartlett and Prisbrey, 1997; Bartlett, 1997).

6.7. SELECTED EXAMPLES OF CHARACTERIZING GOLD TAILINGS

6.7.1. Gold in tailings from David Bell circuit, Teck-Corona, Hemlo, Ontario

Tailings samples from the David Bell circuit of Teck-Corona mine in Hemlo, Ontario, containing 0.5 to 1.6 g/t Au, were analysed to determine the mode of occurrence of the gold in the tailings (Pinard and Petruk, 1991; Laflamme and Petruk, 1991). The tailing was first leached with HF to dissolve the silicate minerals, and then with HNO_3 to dissolve the sulfides. The residue contained a large corroded gold grain (Pinard and Petruk, 1991) which indicated that some of the Au in the tailings was present as incompletely cyanidated grains. In a second investigation sieved fractions were separated into sink and float sub-fractions to concentrate the gold minerals, and polished sections of the sink fraction were studied by optical microscopy, scanning electron microscope and electron microprobe (Laflamme and Petruk, 1991). It was found that the gold was present as (1) liberated and exposed aurostibite ($AuSb_2$) grains that range from 10 to 50 μm in size, (2) minute grains of Au-Hg-Ag alloy encapsulated in silicates and pyrite, and (3) a large liberated gold grain that had not been cyanided. It was interpreted that the aurostibite and encapsulated gold were the major gold carriers in the tails and accounted for most of the gold loss, since they are insoluble in cyanide solutions..

Other studies by Chryssoulis and Winckers (1996) using leaching test, SIMS, LIMS and microprobe analyses showed that the liberated gold grains (< 20 μm) were significant carriers of gold. Chryssoulis and Winckers (1996) found that the liberated gold grains were residual grains which did not dissolve in the cyanide solution because they were coated with surface contaminants including S, Hg, As, Sb and Ag. They showed by LIMS analysis that, prior to cyanidation the major surface element on the gold particles was Au and the surface contaminants were Fe, Ca, K, Na, Al and C. After cyanidation the Au peak was no longer the dominant peak because of surface contamination. Some of the contaminants (Mg, Ca, Fe) had accumulated with time and passively blocked the gold particle surfaces, whereas others (S, Hg, Ag, As and Sb) played a more active role. Chryssoulis and Winckers (1996) reported the composition of the gold alloy as 78.9 wt % Au, 18.5 wt % Hg and 2.6 wt % Ag, and Harris (1989) reported the average composition of the gold alloy in the David Bell orebody as 83.6 wt % Au, 11.6 wt % Hg and 4.6 wt % Ag. It is interpreted that some of the Hg released by dissolution of the gold alloy contaminated the surfaces of the gold alloy particle.

6.7.2. Gold tailings, Nor Acme Mine, Snow Lake, Manitoba

A laboratory mineral processing test had been conducted on material from an old tailings pile, which contained about 5 g/t Au, to determine whether the gold can be recovered. Only 31.4 %

of the gold was recovered in the laboratory test which recovered most of the arsenic (87.9 %) (Table 6.2) and arsenopyrite (84.3 %).

Table 6.2
Assays and distributions of gold in Nor-Acme tails

	Assays		Distributions	
	Au (ppm)	As (wt %)	Au (%)	As (%)
Feed	5	0.6	100	100
Concentrate	28.5	8.8	31.4	87.9
Tail	3.7	0.07	68.6	12.1

Samples of the feed, concentrate and tail were analysed to find the reason for the Au loss. The samples were sieved into sub-fractions and polished sections were prepared from the sieved sub-fractions. A gold search was conducted on each polished section with the image analyser. Six gold grains were found in the feed sample, forty in the concentrate and thirteen in the tail. The data were processed by a materials balancing program to calculate the distribution of gold among the different types of gold occurrences (Table 6.3).

Table 6.3 shows that 40 % of the gold is associated with gangue minerals (23 % exposed and 17 % encapsulated), but only 13 % was recovered (10 % exposed and 3 % encapsulated) (column 2). Similarly 37 % of the gold occurs as free gold (17 % smaller than 3 μm), but only 7 % was recovered in the concentrate. In contrast, only 23 % of the gold is associated with arsenopyrite (18 % exposed and 5 % encapsulated), but only 11 % was recovered (8 % exposed and 3 % encapsulated). The flotation test recovered 84.3 % of the arsenopyrite, but apparently the test was not designed to float gold since much of the exposed gold (even on arsenopyrite) was rejected.

Table 6.3
Occurrence and Distribution of Native Gold in Nor-Acme Tails

Mode of Au occurrence	Au Dist. (feed) (wt units)	Au Dist. (conc) (wt units)	Au Recovery (% in conc)
Au exposed, arsenopyrite	18	8	43
Au encapsulated, arsenopyrite	5	3	76
Au exposed, gangue	23	10	44
Au encapsulated, gangue	17	3	16
Au sulfides	0.4	0.4	100
Au free <3 μm	17	2	14
Au free	20	5	24
TOTAL	100	31	31

CHAPTER 7

APPLIED MINERALOGY: PORPHYRY COPPER DEPOSITS

7.1. INTRODUCTION

Applied mineralogy is an essential tool in exploration of porphyry copper deposits and in mineral processing of porphyry copper ores. It is used to a greater or lesser degree throughout the entire period of exploitation of most deposits. An important aspect of exploration of porphyry copper deposits is characterizing the host rocks and mineralized zones, and classifying the deposits. The characterization involves (1) identifying the minerals, (2) determining mineral quantities, and (3) determining the textural relationships between the minerals, mineralized veins and the wallrock. The main minerals are normally identified by examining hand specimens or drill cores, but the identities of some minerals need to be determined or confirmed by other techniques; usually by microscopical studies of thin or polished-thin sections, supplemented by XRD, SEM/EDX and microprobe (MP) analyses. Textural relations of minerals are determined by optical microscopy and/or SEM/EDX. The samples should be examined by a competent petrographer, as no field geologist looking at thin sections part time can acquire the breadth of experience of a person dedicated to petrography. Furthermore, the petrographer needs feedback from the field geologist to provide usable information related to exploration rather than just rock descriptions (Williams and Forrester, 1995) . The petrographer in turn needs to know the characteristics of porphyry copper deposits to recognize the different features.

Applied mineralogy related to mineral processing of porphyry copper ores involves:
- identifying the minerals,
- determining mineral quantities,
- determining the grain sizes and liberations of the ore minerals,
- determining the minor and trace element contents (e.g. Ag, Au, Se, Te, Ge) of some ore minerals, particularly the chalcopyrite, bornite and pyrite,
- determining the identities and quantities of clay minerals.

7.2. CHARACTERISTICS OF PORPHYRY COPPER DEPOSITS

Porphyry copper deposits are a diverse group of large tonnage, low-grade deposits with a metal assemblage that includes some or all of copper, molybdenum, gold and silver. The deposits are closely associated with, and related to, emplacement of intermediate to felsic, hypabyssal and porphyritic intrusions (Sutherland Brown and Cathro, 1976; Titley and Beane, 1981; McMillan et al., 1995). The mineralization consists of quartz, ore minerals, and silicate minerals in fracture stockworks, veins, vienlets and disseminations in the wallrock (Sikka and Nehru, 1997). The ore zone is silicified and is associated with zones of hydrothermal alteration. The extent of alteration, and the coincidence of alteration zones with ore zones, is variable from one deposit to the next. The degree of zonation is influenced by maturity, sulfur content and depth of porphyry copper deposits (Williams and Forrester, 1995). Another important aspect is that most porphyry copper deposits are capped by a zone of supergene alteration that is generally enriched in Cu and Au.

Two accessory minerals, rutile and apatite with distinctive features, have been found to be characteristic of porphyry copper deposits (Williams and Forrester, 1995). The porphyry copper rutile has a low c/a crystallographic ratio and is red due to a copper content of up to 600 ppm. The porphyry copper apatite occurs as poikalitic crystals, overgrowths and corrosions on other minerals, and fluoresces bright orange. The fluorescence intensity increases with increasing development of a porphyry copper system, but the fluorescence is low in low grade porphyry copper deposits.

7.2.1. Primary ore minerals

The primary ore minerals occur as irregular grains and masses in the quartz veins, and as veinlets and disseminations in the wallrock. The minerals are distributed in zones about the center of the orebodies, but the extent of zonation is variable and some of the characteristic zones may be absent in some deposits. In particular, bornite-rich zones commonly occur at the center of the orebody, but the bornite zone is absent in many orebodies. The distribution of mineralization, from the center to the periphery of the orebodies, is generally bornite-chalcopyrite, chalcopyrite-bornite-pyrite, chalcopyrite-pyrite and pyrite (Waldner et al., 1976; McMillan, 1976b). The chalcopyrite-pyrite zone usually forms the bulk of the deposit. Molybdenum, enargite, and gold are frequently associated with chalcopyrite, but the gold content is low in deposits that contain molybdenite. The depositional sequence observed in the Malanjkhand in India (Bhargava and Pal, 1999), is magnetite - pyrite - chalcopyrite. Late veinlets, vugs, and disseminations of tetrahedrite, tennantite, boulangerite, sphalerite, galena, geocronite, wittichenite, gypsum and zeolites occur in many deposits. The ore minerals that have been found in porphyry copper deposits (Pilcher and Mcdougall, 1976; Johnson, 1973) are listed in Table 7.1.

7.2.1.1. Vein density
As the bulk of the ore minerals occurs in veins, the density of veins has been widely studied. In particular, Oriel (1972) reported that the density of the veins in the Brenda deposit in British Columbia, Canada, ranged from <9 per meter near the periphery of the orebody to 63 per meter and occasionally 90 per meter near the center of the orebody (Soregaroli and Whitford, 1976). The veins are complex, as they may show characteristics of fracture filling in one part and of replacement in another.

7.2.1.2. Silicification
Silicification is an important characteristic of porphyry copper deposits, as quartz is the predominant constituent of the stockworks, veins and veinlets, and of the country rock. Most of the quartz is in fracture controlled veins, although some occurs in the rock as overgrowths on primary quartz, and some locally replaces the country rock. Some of the quartz is hydrothermal and some was produced by alteration of the silicate minerals, particularly plagioclase.

Table 7.1.
Ore minerals found in porphyry copper deposits

Major		Minor		Trace	
Chalcopyrite	$CuFeS_2$	Pyrrhotite	$Fe_{1-x}S$	Gold	Au
Bornite	Cu_5FeS_4	Magnetite	Fe_3O_4	Sphalerite	ZnS
Pyrite	FeS_2	Molybdenite	MoS_2	Galena	PbS
		Chalcocite	Cu_2S	Gudmundite	FeSbS
		Digenite	Cu_9S_5*	Arsenopyrite	FeAsS
		Tetrahedrite	$(Cu,Fe)_{12}Sb_4S_{13}$	Cobaltite	CoAsS
		Tennantite	$(Cu,Fe)_{12}As_4S_{13}$	Geocronite	Pb_5SbAsS_8
		Enargite	Cu_3AsS_4	Boulangerite	$Pb_5Sb_4S_{11}$
				Cosalite	$Pb_2Bi_2S_5$
				Bismuthinite	Bi_2S_3
				Wittichenite	$Cu_3 BiS_3$
				Cubanite	$CuFe_2S_3$
				Mackinawite	$(Fe,Ni)_9S_8$
				Carrollite	$Cu(Co,Ni)_2S_4$
				Silver	Ag
				Hematite and specularite	Fe_2O_3
				Cassiterite	SnO_2
				Ilmenite	$FeTiO_3$
				Rutile	TiO_2
				Leucoxene (altered ilmenite)	
				Powellite	$CaMoO_4$
				Scheelite	$CaWO_4$
				Wolframite	$(Fe,Mn)WO_4$

* Some of the reported digenite (Cu_9S_5) in porphyry copper deposits may be anilite (Cu_7S_4), particularly if the mineral was not X-rayed (Sikka, 2000 - personal communication).

7.2.2. Alteration

Hydrothermal alteration has produced mineral assemblages that can be broadly classified into zones, although a distinct boundary between the zones is seldom apparent. The alteration zones from the center to the outer edges of the deposit, usually follow the sequence: potassic - phyllic - argillic - propylitic (Lowell and guilbert, 1970; McMillan, 1976a). In some deposits propylitic alteration occurs after the potassic alteration, as well as at the outer edge. The potassic alteration has produced biotite and/or K-felspar zones that generally coincide with the center of the orebody. In the Morrison mine a chlorite-carbonate zone surrounds the biotite zone. The phyllic zone contains sericite and quartz and is generally referred to as a sericite-quartz zone. The clay minerals of the argillic zone form a clay minerals zone. The clay minerals zone and the quartz-sericite zone frequently correspond with the main part of the ore mineral zone. Epidote occurs in propylitic alteration zones that are generally at the outer margins of orebodies. A magnetite zone is present at the periphery of the Afton orebody in British Columbia, Canada, and is flanked by a barren pyrite zone (Carr and Reed, 1976).

7.2.2.1. Potassic alteration

Potassic alteration is characterized by hydrothermal biotite and/or K-feldspar alteration along quartz-sericite veins and in fracture veins, usually at the central part of the ore deposit. At the Valley Copper deposit in British Columbia, Canada, K-felspar alteration is associated with vein sericite alteration where K-feldspar replaces sericitized plagioclase or vein sericite (Figure 7.1) (Osatenko and Jones, 1976). Similarly, a zone of K-feldspar alteration has been reported in the Malanjkhand deposit in India (Sarkar et al., 1996; Bhargava and Pal, 1999). In the Morrison and Bell Copper mines in British Columbia, Canada, the ore zones are within centrally located biotite zones (Carson et al., 1976; Carson and Jambor, 1976). The hydrothermal biotite (Carson et al., 1976; Bhargava and Pal, 1999) has replaced primary mafic minerals in the surrounding rocks, and at the periphery of the biotite zone, it is brown and green and is accompanied by chlorite. As the orebody is approached the biotite becomes a deeper brown, coarser-grained and chlorite diminishes. In the El Teniente mine in Chile both K-felspar and biotite are the alteration minerals in the main bornite-chalcopyrite zone.

7.2.2.2. Phyllic alteration

Phyllic alteration is characterized by envelopes of quartz and flaky sericite on both sides of quartz veins and mineralized veins in zones of intense fracture veins (Figure 7.2). The phyllic alteration is commonly developed throughout the ore zone. At the Lornex deposit in British Columbia, Canada, the phyllic zone extends into the argillic zone as a grey mixture of quartz and sericite borders on quartz-copper sulfide and quartz-molybdenite veins (Waldner et al., 1976). The sericite is commonly associated with small amounts of kaolinite and montmorillonite and occasionally with calcite and epidote.

Figure 7.1. Potassic alteration: K-feldspar (K, white) along edges of quartz veinlet, and associated with sericite-quartz (ser-qtz), (modified from Osatenko and Jones, 1976).

Figure 7.2. Phyllic alteration: Sericite-quartz envelopes along quartz veinlets (qtz), (modified from Osatenko and Jones, 1976).

7.2.2.3. Argillic alteration

The term argillic alteration is used to describe the alteration of feldspars, and locally mafic minerals, to an assemblage of sericite, kaolinite (with or without montmorillonite), and minor chlorite. This type of alteration occurs within the ore deposits and often extends beyond the minable grade isopleth (McMillan, 1976a). At the Lornex deposit the cores of plagioclase crystals are more intensely altered than the rims, whereas alteration of orthoclase progresses from the outside towards the core (Waldner et al., 1976). The orthoclase alters to kaolinite, sericite and minor montmorillonite, and biotite and hornblende alter to chlorite and sericite. At Valley Copper (Osatenko and Jones, 1976) the plagioclase has been completely altered to a mixture of sericite, kaolinite, quartz and calcite, and biotite has been completely altered to sericite, siderite, kaolinite and quartz. At the Highmont deposit in British Columbia, Canada, the plagioclase has been extensively replaced by kaolinite, montmorillonite and a carbonate (Reed and Jambor, 1976). In zones of intense alteration and along faults in the Morrison mine, biotite, hornblende and plagioclase phenocrysts have been almost totally altered to kaolinite (with or without montmorillonite), chlorite and mixtures of calcite, dolomite and rarely siderite. Williams and Forrester (1995) stated that kaolinite may not be as abundant in the argillic zone as generally reported, because much of the sericite may have been mis-identified as kaolinite.

7.2.2.4. Propylitic alteration

The argillic alteration zone generally grades into the propylitic alteration zones and in places the two zones overlap. Epidote is commonly used as the characteristic mineral of propylitic alteration. Feldspars in propylitic zones are altered to sericite, carbonate (calcite) and some clay minerals, and mafic minerals are altered to chlorite, carbonates, sericite and epidote. At the J.A. and Highmont deposits in the Highland Valley in British Columbia propylitic facies assemblages occur almost throughout the ore zones (McMillan, 1976b; Reed and Jambor, 1976). At the Lornex deposit there is a narrow propylitic alteration zone peripheral to the orebody (Waldner et al., 1976). The assemblage consists of epidote, chlorite, and carbonates (calcite) with minor sericite and hematite. Epidote and calcite commonly occur as veins. At the Valley Copper deposit propylitic alteration is weakly developed. The magnetite is commonly altered to hematite (*var.* specularite).

7.2.2.5. Distribution of other minerals

Gypsum and anhydrite are present in small amounts as veinlets, some mineralized, and as disseminated grains in the rock. Large gypsum crystals have been found in late mineralized veins in the El Teniente mine.

Zeolites are commonly intergrown with calcite, and locally with sulfide minerals, gypsum and epidote. They are younger than the main minerals in the deposits, and occur as veins, fracture coatings, and as disseminated grains in pervasive alteration zones around fractures and zeolite veins. The main zeolite is laumontite, but some stilbite and uncommon chabazite and heulandite occur.

Tourmaline commonly occurs in and near breccia bodies where it is present as crystalline aggregates, and as replacements of clasts and fractured rock. The mineral also occurs as aggregates of grains disseminated in the wallrock. It is commonly associated with quartz, specularite (hematite), epidote, calcite, chalcopyrite, bornite, digenite and occasionally with rare minerals such as actinolite.

7.2.2.6. Hypogene and alteration non-metallic minerals

Main non-metallic hypogene and alteration minerals in porphyry copper deposits

Quartz	SiO_2
K-feldspar	$KAlSi_3O_8$
Plagioclase	$(Na,Ca)Al(Al,Si)Si_2O_8$
Sericite	$KAl_2Si_3O_{10}(OH)_2$
Biotite	$K(Mg,Fe)_3(Al,Fe)Si_3O_{10}(OH,F)_2$
Hornblende	$Ca_2(Mg,Fe)_4Al(Si_7Al)O_{22}(OH,F)_2$
Kaolinite	$Al_2Si_2O_5(OH)_4$
Chlorite	$(Mg,Fe)_6Al_2Si_2O_{10}(OH)_8$
Epidote	$Ca_2Al_3(SiO_4)_3(OH)$
Calcite	$CaCO_3$

Minor non-metallic hypogene and alteration minerals in porphyry copper deposits

Montmorillonite		$(Na,Ca)_{0.33}(Al,Mg)_2Si_4O_{10}(OH)_2.nH_2O$
Tourmaline		$(Ca,K,Na)(Al,Fe,Mg,Mn,Li)_3(Al,Cr,Fe,V)_6(BO_3)_3Si_6O_{18}(OH,F)_4$
Gypsum		$CaSO_4.2H_2O$
Anhydrite		$CaSO_4$
Albite		$NaAlSi_3O_8$
Zeolites:	*laumontite*	$CaAl_2Si_4O_{12}.4H_2O$
	stilbite	$NaCa_2Al_5Si_{13}O_{36}.14H_2O$
	chabazite	$CaAl_2Si_4O_{12}.6H_2O$
	heulandite	$(Na,Ca)_{2-3}Al_3(Al,Si)_2Si_{13}O_{36}.12H_2O)$

Trace non-metallic minerals in porphyry copper deposits

Actinolite	$Ca_2(Mg,Fe)_5Si_8O_{22}(OH)_2$
Prehnite	$Ca_2Al_2Si_3O_{10}(OH)_2$
Alunite	$KAl_3(SO_4)_2(OH)_6$
Jarosite	$KFe_3(SO_4)_2(OH)_6$
Monazite	$(La,Ce,Nd,Th)PO_4$
Apatite	$Ca_5(PO_4)_3F$
Zircon	$ZrSiO_4.$
Corundum	Al_2O_3
Dolomite	$CaMg(CO_3)_2$
Siderite	$FeCO_3$

7.2.3. Maturity

Maturity is defined as the degree of magmatic differentiation before crystallization (Williams and Forrester, 1995), and is characterized on the basis of (1) rock type, (2) alteration and mineralization processes, and (3) content and relationships of deuteric mineral assemblages. Three rock types are typically recognized in porphyry copper systems: tonalite, granodiorite and quartz monzonite, with maturity increasing from tonalite to quartz monzonite.

The alteration and mineralization is developed to a maximum degree in mature porphyry copper deposits. The quartz-sericite-pyrite zone (phyllic alteration zone) in mature systems is large and the secondary biotite zone is small. Furthermore, the best copper values are somewhere below the top of the system, commonly at the boundary between the phyllic and potassic alteration zones. Hence if a porphyry copper intrusion is mature, a grade of 0.3 % Cu in the phyllic alteration zone could increase to more than 1 % Cu at the boundary between the phyllic and potassic alteration zones. In contrast, the grade in an immature porphyry copper system will not increase towards the potassic alteration zone.

Deuteric water is tied up in calcium silicates, chlorite, hydrobiotite and epidote in immature porphyry copper intrusions. In contrast, deuteric veinlets in mature intrusions have sericite selvages. Immature porphyry copper systems, especially those associated with tonalite intrusions, commonly have higher gold and silver contents and higher bornite:chalcopyrite ratios than equivalent mature porphyry copper systems (Williams and Forrester, 1995).

7.2.4. Role of sulfur

Porphyry copper deposits contain up to 8 % sulfur. Williams and Forrester (1995) reported that deposits with high sulfur contents (e.g. high chalcopyrite + pyrite contents) tend to (1) correlate with greater maturity, (2) occur at higher crustal levels, and (3) have broad quartz-sericite alteration aureoles. Williams and Forrester (1995) suggest that the sulfur promotes destruction of the crystal structure of mafic silicate minerals and captures the Fe. The K, Si, Al and H_2O are released and are available to form sericite. The sulfur, therefore, plays a role in the development of the hydrothermal alteration zones referred to above.

7.2.5. Depth of porphyry copper system

An ideal porphyry copper deposit has a vertical range of about 2,450 m (~8,000 feet). The rock at the bottom of the system is fresh, and only narrow late magmatic quartz veinlets that may contain traces of molybdenite and bornite are present. The copper content of the rock at depth varies between 500 and 2,000 ppm with most of the copper occurring as tiny threads of chalcopyrite along cleavage planes in biotite (Williams and Forrester, 1995).

The first diagnostic features of a porphyry copper system appear at 2,100 to 1,500 m (7,000 to 5,000 feet) below the top of the system. The rock is still fresh, but there is a development of pleochroic blue corundum commonly rimmed by muscovite in the core of fresh-looking calcic plagioclase crystals. The quartz-corundum assemblage is unstable and reacts to form orthoclase. The quartz veins at this level are irregular and contain orthoclase and chalcopyrite.

At 900 to 1,500 m (3,000 to 5,000 feet) below the top of the system the classic Lowell and Guilbert (1970) alteration zoning develops. The quartz veins become wider and are abundant, and the copper content increases.

In the top 900 m of the system the rock textures are destroyed and sericite replaces virtually everything but primary quartz, forming the quartz-sericite (phyllic) alteration zone. It is noteworthy that at the Malanjkand deposit in India (Bhargava and Pal, 1999) the copper mineralization decreases around 700 m below the surface of the orebody.

7.2.6. Supergene mineralization

A zone of supergene mineralization is present at top of most porphyry copper deposits. It consists of an upper oxidized zone called a cap, and an underlying sheet-like zone commonly referred to as a blanket (Ney et al., 1976). The cap generally contains a variety of supergene copper minerals and may be enriched in Au. However, in a few deposits, the cap does not contain copper minerals because the elements released from the Cu sulfides were washed down and precipitated at the top of the blanket zone. The supergene copper minerals in the blanket zone are Cu sulfides that form below the water table. The lower margin of the supergene blanket grades into the underlying hypogene Cu sulfide mineralization.

7.2.6.1. Mineralogy of cap

The ore at the top of some ore bodies has been intensely oxidized and the ore minerals were dissolved and precipitated in-situ as supergene minerals. The supergene copper minerals in the cap include Cu oxides, Cu carbonates, Cu sulfates, Cu chlorides, Cu silicates and native copper (Table 7.2).

The Cu oxides include tenorite, cuprite, delafossite and copper bearing goethite. The cuprite in the Malanjkhand deposit in India occurs in veins that are up several cm wide. It is associated with native copper, goethite and delafossite, contains inclusions of Cu chlorides and native copper, and is generally surrounded by massive goethite (Petruk and Sikka, 1987). Delafossite occurs near the cuprite-goethite interface as large irregular grains in goethite and as minute inclusions in cuprite. Goethite is ubiquitous in the oxidized ore. The goethite associated with cuprite generally surrounds the mineral as stringers and can contain up to 1.33 % Cu (Petruk and Sikka, 1987).

The Cu carbonates are malachite and azurite. The malachite is much more abundant than azurite. It occurs as intensely oxidized material and as encrustations on cuprite and goethite. The malachite-bearing material is so intensely oxidized that it has lost the features of the primary ore. Associated plagioclase is generally altered to clay minerals and only the quartz and remnants of other minerals, including brochantite, are preserved. The malachite is typically present as masses with remnants of other minerals. Azurite is associated with malachite and occurs as irregular grains, veinlets and encrustations in cavities.

The Cu sulfates and silicates are brochantite, conichalcite, chalcanthite and chrysocolla, and are associated with jarosite and alunite. Brochantite is present as replacements of chalcocite and as inclusions in malachite. Alunite occurs in intensely oxidized ore and is present as large grains associated with gibbsite, chlorite and azurite.

The Cu chlorides occur as inclusions in cuprite. The Cu chlorides found in the cuprite from the Malanjkhand deposit are nantokite, paratacamite and claringbullite (Petruk and Sikka, 1987).

Native copper occurs as veins and veinlets in the rock and as inclusions in cuprite. Some of the native copper in veinlets is dendritic and some is vuggy. In the Afton deposit native copper forms between 65 and 85 % of the copper minerals in the oxidized zone (supergene cap) and occurs as dendrites, films, granules and masses up to 5 mm in size (Carr and Reed, 1976).

Table 7.2
Supergene minerals in cap and blanket zones

Native copper		Cu sulfides	
Copper	Cu	Bornite	Cu_5FeS_4
		Covellite	CuS
Cu oxides		Chalcocite	Cu_2S
Tenorite	CuO	Digenite	Cu_9S_5
Cuprite	Cu_2O	Idaite	Cu_3FeS_4
Delafossite	$CuFeO_2$	Covellite	
		Yarrowite	$Cu_{1.05-1.14}S$
Cu carbonates		Spionkopite	$Cu_{1.27-1.45}S$
Malachite	$Cu_2(CO_3)(OH)_2$	Geerite	$Cu_{1.6}S$
Azurite	$Cu_3(CO_3)_2(OH)_2$	Anilite	Cu_7S_4
		Djurleite	$Cu_{31}S_{16}$
Cu sulfates and silicates			
Brochantite	$Cu_4(SO_4)(OH)_6$	Other supergene minerals	
Conichalcite	$CaCu(AsO_4)(OH)$	Gibbsite	$Al(OH)_3$
Chalcanthite	$CuSO_4.5H_2O$	Pyrrhotite	$Fe_{1-x}S$
Chrysocolla	$(Cu,Al)_2H_2Si_2O_5(OH)_4.nH_2O$	Marcasite	FeS_2
Jarosite	$KFe_3(SO_4)_2(OH)_6$	Ferrimolybdite	$Fe(MoO_4)_3.3H_2O$
Alunite	$KAl_3(SO_4)_2(OH)_6$	Goethite	$FeO(OH)$

Cu chlorides	
Nantokite	CuCl
Paratacamite	$Cu_2(OH)_3Cl$
Claringbullite	$Cu_4Cl(OH)_7.1/2H_2O$

7.2.6.2. Mineralogy of blanket zone

The ore minerals in the blanket zone are secondary copper and iron sulfides. The degree of oxidation in the blanket zone varies from minor at the interface between the primary and supergene ore where the chalcopyrite and primary bornite are only slightly altered and tend to be bordered by a rim of covellite, chalcocite or digenite, to intensely altered at the interface between the blanket zone and supergene cap. The main copper minerals in the blanket zone are chalcocite, covellite, bornite, remnants of unaltered chalcopyrite, and minor to trace amounts of some, or all, of the minerals of Cu sulfide series minerals which alter in the forward and reverse sequence bornite - idaite - covellite(yarrowite, spionkopite) - geerite - anilite - djurleite - chalcocite (sikka et al.,1991). Chalcocite is an abundant mineral in the blanket zone and occurs as disseminations and in veins, some up to 25 mm wide, as in the Afton deposit (Carr and Reed, 1976). The pyrite of the primary ore alters to pyrrhotite, marcasite and iron oxide phases. The molybdenite alters to ferrimolybdite.

7.3. APPLIED MINERALOGY RELATED TO MINERAL PROCESSING

Since porphyry copper ores are generally composed of a cap (intensely oxidized ore), a blanket zone (moderately oxidized ore), and primary ore (Sikka and Bhappu, 1992), different mineral processing techniques are required to recover the copper from the different ore types. The amount of copper contained in intensely oxidized and partly oxidized ores is frequently enough to install separate flowsheets or circuits for each ore type.

7.3.1. Mineralogical characterization of intensely oxidized ores

The copper is commonly recovered from the intensely oxidized zones by leaching with acid. Gold, if present, is recovered by cyanidation. A characterization, by mineralogical techniques, of intensely oxidized ores and leach residues, prior to and/or during leaching, helps the operators design and maintain an efficient leaching process. The mineralogical characterization would determine the identities, sizes and quantities of the minerals (i.e. Cu oxides, Cu carbonates, Cu sulfates, Cu silicates, gangue minerals and residual sulfides), and degree of exposure of the Cu bearing minerals to the leaching solutions. The techniques and instruments for conducting the mineralogical characterization studies are described in Chapters 1 and 2. The mineralogical information would:
- provide a guideline for determining (a) acid strength and (b) leach time,
- identify gangue minerals that are soluble in the acid. This would qualitatively indicate:
    - (a) acid consumption,
    - (b) whether some minerals would release elements that might interfere with subsequent precipitation of the copper from solution,
- identify minerals that would not dissolve in the acidic solution (e.g. silicates, pyrite),
- identify Cu-bearing minerals which are not soluble in acidic solutions (e.g. chalcopyrite).

7.3.2. Mineralogical characterization of partly oxidized ores

The partly oxidized ores that occur in the blanket zone contain secondary Cu sulfides. The secondary Cu sulfides release large amounts of Cu ions into solution during flotation. Consequently, sulfides such as pyrite, arsenopyrite, sphalerite, galena, etc. are activated by the Cu ions and pulled into the rougher copper concentrate. Subsequent cleaning may reject some of these impurities, but it will also reject some chalcopyrite and bornite with a high copper loss. On the other hand, the cleaning might not reject enough unwanted minerals, such as arsenic-bearing minerals. Furthermore, the large amount of Cu ions in solution may alter the reagent concentration and affect flotation of chalcopyrite and bornite.

It may be possible to obtain better recoveries of copper from the partly oxidized ores, and to obtain better rejection of the deleterious minerals by installing an oxides circuit that recovers oxidized copper minerals (chalcocite, covellite, cuprite, etc.) (Maurice, 1979; De Cuyper, 1977; Castro et al., 1974). The oxides circuit would be installed at the front end of the circuit that processes primary ore. The oxides circuit would remove the secondary Cu sulfides and reject chalcopyrite, bornite and the other sulfides to the oxides circuit tailings. The tailings from the oxides circuit would be reconditioned and processed in the primary circuit to recover the chalcopyrite and bornite, and to reject minerals such as arsenopyrite, gudmundite and cobaltite.

A mineralogical study in connection with mineral processing of partly oxidized ores would involve:

- identifying the secondary and primary sulfides,
- determining the liberations and sizes of the secondary and primary Cu sulfides,
- determining the associations of the unliberated Cu sulfides.

The liberation measurements would report:

- the proportion of secondary and primary Cu-bearing minerals that are liberated,
- the proportion of unliberated Cu-bearing minerals in recoverable types of particles (e.g. particles containing more than 50% Cu-bearing minerals, and are larger than 10 μm),
- the proportion unliberated Cu-bearing minerals that would be recoverable after regrinding (particles containing >20% and <50% Cu-bearing minerals, and are larger than 37 μm),
- proportion unliberated and unrecoverable Cu-bearing minerals (particles containing <20% Cu-bearing minerals and are small).

The mineral association data are generally reported as proportion of mineral occurring as unliberated grains in different types of particles (e.g. in pyrite-rich particles, gangue-rich particles, etc.). Mineralogical data for use by mineral processing engineers must be concise and reported in a standard format of one page per sample.

7.3.3. Mineralogical characterization of primary ores

Much of the chalcopyrite and bornite in porphyry copper ores is relatively coarse-grained, therefore, a high recovery is generally obtained by mineral processing if the ore does not contain secondary Cu sulfide and Cu oxide minerals that interfere with the chalcopyrite-bornite flotation. On the other hand, inevitably, some of the ore is fine-grained and the copper and molybdenum sulfides may not be fully liberated. The unliberated minerals could also be associated with minerals that are difficult to process. Regrinding may, therefore, be required to liberate the minerals, particularly molybdenite. Other aspects such as floating of As-bearing minerals, the deportment of silver and gold, the occurrence of selenium and tellurium, and the distribution of clay minerals, are factors that may need to be considered and studied.

Mineralogical studies are commonly performed at the minesite on polished sections of monthly or weekly composite samples to determine:

- identities of sulfide and oxide minerals,
- quantities of sulfide and oxide minerals,
- liberation characteristics of the Cu-bearing minerals,
- sizes of unliberated Cu-bearing minerals,
- associations of unliberated Cu-bearing minerals.

The minerals are usually identified by optical microscopy studies. Mineral quantities, liberation analysis and mineral associations are frequently determined by point counting, but more accurate data would be obtained by image analysis. It is likely that in the future point counting will be replaced by image analysis, either at the minesite or contracted out.

The liberation analyses and measurements of mineral associations in primary ores are performed and reported in the same manner as described above for partly oxidized ores.

Studies that need to be performed in co-ordination with mineral processing include determining the distribution of arsenic, selenium, tellurium, silver, and gold among the ore minerals, especially the Cu-bearing minerals. The distribution of arsenic (As) needs to be determined because it is a hazardous element, particularly in smelter dusts. In contrast, some of the other elements may be valuable by-products that, if recovered, may enhance the value of the ore.

Determining the distribution of arsenic involves identifying all the arsenic bearing minerals in the feed, rougher concentrate and tailings in a nominal size fraction from a mineral processing operation, for example 65 to 400 mesh fraction (212 to 37.5 μm). The potential As-bearing minerals are enargite, tennantite, arsenopyrite, cobaltite, gudmundite and pyrite. The quantity of arsenic in pyrite is determined by either a microprobe or PIXE. The quantity of each As-bearing mineral is determined by either image analysis or point counting. The distribution of arsenic among the various minerals is calculated, and a decision is made about rejecting specific minerals. If the ore contains a small amount of arsenic, the study is performed by pre-concentrating the As-bearing minerals with a heavy liquid having a specific gravity of 2.96. The sink and float sub-fractions are assayed for arsenic to determine the distributions of arsenic, and polished sections of the sink sub-fractions are analyzed by either image analysis or point counting to determine the quantity or proportion of each As-bearing mineral.

The distribution of Se, Te and Ag among the ore minerals is determined by using a technique that is similar to determining the distribution of arsenic, but the primary ore minerals (chalcopyrite, bornite and pyrite) and secondary Cu sulfides have to be analyzed by PIXE to determine the trace quantities of these elements in the minerals. The quantity of each mineral in the sample is determined by image analysis. The distribution of the elements among the minerals is calculated.

It has been observed at many concentrating operations that clay minerals, particularly montmorillonite, have a detrimental effect on flotation of primary and secondary copper minerals. Gorodetskii et al. (1973) have shown that slimed clay minerals, especially montmorillonite, reduce recoveries of copper and molybdenum sulfides by absorbing and entrapping water and reagents and increasing the viscosity of the pulp. Furthermore, recoveries decrease as the fineness of grind of the clay-rich material increases. Recoveries are improved by desliming, particularly if desliming removes the clay minerals. Information on identities and quantities of clay minerals is therefore essential.

7.4. IDENTIFYING MINERALS

Identification of ore minerals that are grey in polished sections is extremely difficult. In particular, the minerals chalcocite, digenite, tennantite, tetrahedrite, freibergite, tenorite and enargite have similar shades of grey in reflected light. These minerals cannot normally be differentiated, even by a skilled and experienced microscopist, without additional analyses such as XRD or SEM/EDX. Therefore, their quantities cannot be determined by point counting or with an image analyzer interfaced to an optical microscope. On the other hand, these minerals have slightly different average atomic numbers, ranging from 23.5 to 37.3 (Table 7.3), hence it is possible to distinguish them using a backscattered electron (BSE) image produced with a SEM

(See Chapter 2) operating at appropriate conditions. The SEM should operate at a relatively high voltage (about 20 kV) and current (about 15 nA), and the gain and suppress should be set to detect minerals with average atomic numbers between 23.5 to 35 (e.g. all minerals with average atomic numbers below 23.5 should appear black and all minerals with atomic numbers above 35 should appear white in the BSE image). If mineral separation is achieved by this technique the relative mineral quantities can be determined with an image analyzer interfaced to a SEM. If, however, the minerals cannot be differentiated in the BSE image, they can be identified with an image analyzer that uses an X-ray technique to determine the relative element contents in the minerals. The image analysis system at CANMET and the QEM*SEM have such capabilities (see Chapter 2).

Table 7.3
Average atomic number of copper-bearing minerals

Mineral	Formula	Atomic No
Chalcopyrite	$CuFeS_2$	23.5
Idaite	Cu_3FeS_4	24.1
Covellite	CuS	24.6
Tenorite	CuO	24.8
Bornite	Cu_5FeS_4	25.3
Enargite	Cu_3AsS_4	25.5
Yarrowite	$Cu_{1.05-1.14}S$	25.6
Spionkopite	$Cu_{1.27-1.45}S$	25.9
Digenite	Cu_9S_5	26.1
Geerite	$Cu_{1.6}S$	26.1
Anilite	Cu_7S_4	26.3
Djurleite	$Cu_{31}S_{16}$	26.3
Chalcocite	Cu_2S	26.4
Tennantite	$(CuFe)_{12}As_4S_{13}$	26.4
Cuprite	Cu_2O	26.7
Tetrahedrite	$(CuFe)_{12}Sb_4S_{13}$	32.5
Freibergite (35% Ag)	$(Cu_{4.6}Ag_{5.4}Fe_{0.8}Zn_{1.2}Sb_4S_{13}$	37.5

CHAPTER 8

MINERALOGICAL CHARACTERISTICS AND PROCESSING OF IRON ORES

8.1. INTRODUCTION

Iron is recovered from a wide variety of iron deposits (Gross, 1965), and is generally exploited by open pit mining as high tonnage operations. Parts of some deposits consist of relatively pure massive hematite and/or magnetite and are sold in sized lumps (<6 inches) as direct shipping ore. Most iron deposits, however, consist of iron formations that contain gangue minerals. The iron-bearing minerals in these deposits need to be concentrated by mineral processing techniques, which depend upon the mineralogical characteristics of the deposits. The concentrates are usually upgraded and pelletized to meet customer specifications.

Mineral characteristics related to exploitation of iron ores from the iron-formations in the Labrador Trough in Newfoundland and Quebec in Canada, and of Minette-type oolitic sedimentary ironstone in the Peace River area in Alberta, Canada are discussed in this chapter as examples of applied mineralogy related to mineral processing of iron ores.

8.2. IRON ORES IN THE LABRADOR TROUGH

The central part of the Labrador Trough contains three areas of iron deposits. They are the Carol Lake deposits of the Iron Ore Company of Canada (IOC) around Labrador City, Newfoundland; the Wabush Mines deposits near Wabush, Newfoundland, about 3 kilometers east of Labrador City; and the Mount Wright deposits near Fermont, Quebec, about 30 kilometers SW of Labrador City.

8.2.1. Characteristics of the ores

The iron-formations consist mainly of hematite, magnetite and quartz, and occur in a stratigraphic succession of metasedimentary rocks, which have been metamorphosed to the epidote-amphibolite and amphibolite facies. The hematite is present as specularite and as irregular grains and clusters of grains in granular quartz. Some of the magnetite occurs as granular grains, and some is recrystallized and is intimately intergrown with quartz and hematite. The quartz has a granular texture that was produced by leaching of silica along the quartz grain boundaries (Gross, 1968). The leaching reduced the intergranular cohesion; hence, the specularite, granular hematite and granular magnetite can be easily separated from the quartz by grinding. In contrast, the recrystallized magnetite does separate as readily from the quartz and hematite. A conspicuous metamorphic feature, particularly in the iron formations, is a marked increase in grain size with increase in metamorphism (Gross, 1968).

Varying amounts of oxidation are present throughout the ores. The oxidation altered the iron carbonates and some hematite to goethite, and converted a minor amount of the magnetite to martite. Some of the iron-bearing carbonates (ankerite and siderite) contain manganese. The oxidation expelled the manganese, which was subsequently precipitated as secondary Mn minerals.

8.2.1.1. Ores of the Iron Ore Company of Canada (IOC), Labrador City, Newfoundland

The ores mined from the various Carol Lake (IOC) deposits contain between 35 and 45 wt % Fe, about 35 to 45 wt % silicate gangue, and low amounts of Mn, P, Al, Ca, Mg and S. They vary from massive ores composed of hematite, magnetite and quartz, to interbanded ores composed of alternating bands of siliceous material, hematite-rich material, and magnetite-rich material. The magnetite and hematite contents are variable throughout the orebodies, but average around 38 wt % hematite and 18 wt % magnetite. Secondary goethite is occasionally present in oxidation zones where it occurs as a powder and as botryoidal and colloform masses. Much of the carbonate is ankerite and siderite that contains significant amounts of Mn.

8.2.1.2. Ores in the Wabush deposits

The ores in the Wabush deposits contain more manganese than the ores in the Carol Lake and Mount Wright deposits. Furthermore, the iron-formation in the Wabush deposits is much more friable and leached than in the Carol Lake deposits, and the decomposed material extends to greater depths. The friability resulted from extensive leaching of silica around quartz grains leaving a minimum of intergranular cohesion. The silica leaching was accompanied by oxidation, which altered the Fe carbonates and some of the hematite to goethite, and converted the magnetite to martite. Another feature associated with the leaching and oxidation is a widespread distribution of secondary Mn oxides in joints, veins and permeable zones (Gross, 1968). An average ore contains about 35 wt % Fe and 1.8 % Mn, and less magnetite than in the Carol Lake (IOC) ores.

8.2.1.3. Ores in the Mount Wright deposits

The ores in the Mount Wright deposits consist of a granular ore composed of quartz and hematite, and of interbanded ores composed of quartz bands and specular hematite bands. The quartz bands vary from about 0.6 to 10 cm in width and contain hematite particles 10 to 100 μm in size. The specular hematite bands are up to 3 mm wide and are composed largely of small masses of hematite, but some of the hematite masses are up to 1 m wide (Gross, 1968).

More than 90 % of the iron in the Mount Wright ores occurs as hematite. Magnetite is present in small amounts, mainly as fine-grained inclusions in hematite. The Fe content of the ores varies between 30 and 35 % (Gross, 1968). Gross reported that the Mount Wright iron-formations have been extensively recrystallized with nearly complete segregation of quartz and iron oxides in coarse grains. Much of the iron-formation is friable and highly decomposed due to intergranular leaching of quartz. Circulating groundwater is considered to be the main agent that removed silica, deposited some goethite, and altered muscovite and feldspar to kaolinite.

8.2.2. Mineralogy

The ore minerals in the IOC deposits are hematite, magnetite, minor goethite, minor siderite and trace manganese-bearing minerals. The Fe-bearing ore minerals in the Wabush deposits are hematite, magnetite, goethite and siderite. Significant amounts of Mn-bearing minerals are also present in the Wabush deposits; they are pyrolusite, psilomelane, wad, Mn goethite, and Mn siderite. The Fe-bearing ore minerals in the Mount Wright deposits are hematite, minor magnetite, trace goethite and trace siderite.

The gangue minerals in the ores of the three deposits are quartz, feldspar, calcite, apatite, amphiboles, pyroxenes, biotite, chlorite, tourmaline, garnet and pyrite. The minerals in the metasedimentary rocks are quartz, feldspar, amphiboles (grunerite, actinolite), pyroxenes (hypersthene, diopside), micas (biotite, phlogopite), chlorite, garnet, carbonates (siderite, ankerite, calcite, dolomite, rhodocrosite), apatite, pyrite and ilmenite;

8.2.2.1. Hematite (Fe$_2$O$_3$)
Hematite is the main ore mineral. It occurs as masses of grains and as discrete grains in granular quartz (Figure 8.1a,8.1b), as specularite where the grains were elongated by metamorphism (Figure 8.2a, 8.2b), and as martite where hematite replaces magnetite. The bonding between most of the hematite and quartz grains is weak due to the leaching of silica around the quartz grains. A trace amount of hematite occurs as minute inclusions in quartz (Figure 8.3), and some contains minute quartz inclusions (Figure 8.4).

8.2.2.2. Magnetite (Fe$_3$O$_4$)
Magnetite occurs as irregular and granular grains, as recrystallized euhedral grains, and as minute inclusions in quartz and hematite. The granular grains occur in massive hematite, massive magnetite and in quartz, and like the hematite, are weakly bonded to the quartz. The recrystallized euhedral grains range from relatively small to large crystals, and are intimately intergrown with the adjacent minerals espcially hematite and quartz. Some of the magnetite in the Wabush deposits, and a minor amount in the IOC deposit is partly to completely replaced by hematite (martitization). Some of the magnetite, particularly in the Wabush deposits, contains Mn; analyses of individual grains reported up to 7.0 wt % Mn.

8.2.2.3. Martite
The term martite is used in the iron industry for secondary hematite that formed by replacing magnetite. The replacement proceeds from the outer edges towards the center of the magnetite grains commonly along crystallographic planes. Some of the martite in the Wabush deposits contains up to 10 wt % Mn.

8.2.2.4. Goethite (FeO(OH))
Goethite is a secondary mineral that formed by oxidation of Fe-carbonates, hematite, martite and magnetite. It occurs as powders and botryoidal masses in open spaces along faults, in interstitial spaces between hematite grains (Figure 8.2a, 8.2b, 8.5), and as an outer layer on hematite and martite grains (Figure 8.6). Goethite masses are present in quartz where the goethite has replaced hematite. Goethite that replaced hematite does not contain Mn, whereas goethite that replaced carbonates and martite contains significant amounts of Mn (up to 27 wt % Mn). The Mn-goethite is microcrystalline and occurs as patches and veinlets in non Mn-goethite and in Mn-rich martite.

152

Figure 8.1a. SEM photomicrograph showing grains and small masses of hematite (white) disseminated in quartz (grey).

Figure 8.1b. Photomicrograph showing disseminated grains of hematite (white) in interstices between gangue minerals (grey).

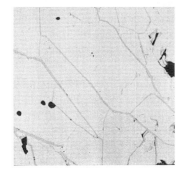

Figure 8.2a. Photomicrograph of specular hematite (light grey) and goethite (dark grey). Wabush deposit.

Figure 8.2b. SEM photomicrograph of specular hematite (light grey) with goethite along grain boundaries. Wabush deposit.

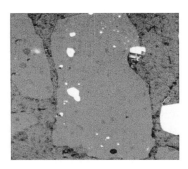

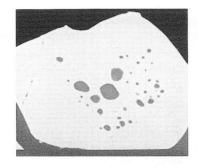

Figure 8.3. Photomicrograph of a polished section of a quartz grain (dark grey) with hematite (white) inclusions.

Figure 8.4. Photomicrograph of a polished section of a hematite grain (white) with quartz (grey) inclusions.

A major problem is encountered in determining the goethite content. Quantitative XRD analysis is a common method of analyzing for phase quantities in fine-grained powders. Unfortunately much of the finely powdered goethite in iron ores is weakly crystalline to non-crystalline, and produces broad XRD peaks that are not proportional to goethite contents. Other techniques, such as thermo-gravimetric analysis are not sensitive enough, and image analysis of finely powdered materials is difficult.

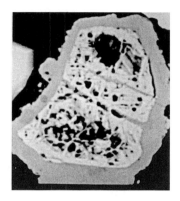

Figure 8.5. SEM photomicrograph of a polished section of hematite (variety specularite) (white) partly replaced by goethite (light grey). Wabush deposit.

Figure 8.6. SEM photomicrograph of a polished section of porous martite (light grey with black zones) partly replaced by goethite (grey) that surrounds the martite. Wabush deposit.

8.2.2.5. Limonite

Some of the goethite in the Wabush deposit has been replaced by a hydrous Fe oxide that contains 25 to 45 wt % Fe and no Mn (Figure 8.7). The identity of this phase has not been determined, although it is commonly referred to as limonite.

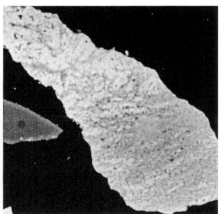

Figure 8.7. SEM photomicrograph of a polished section of a grain of goethite (light grey) partly replaced by limonite (grey). Wabush deposit.

8.2.2.6. Mn Minerals

The Mn minerals are secondary minerals that formed by alteration of carbonates, magnetite and martite, and occur mainly in the Wabush deposit.

8.2.2.6.1. Pyrolusite (MnO_2): Pyrolusite is the main Mn mineral in the Wabush deposits. It occurs as aggregates of crystals associated and intergrown with hematite, Mn-goethite, psilomelane and wad. Some is present in interstices between hematite and magnetite, and some as veinlets in Mn-goethite.

8.2.2.6.2. Psilomelane (Mn oxide with 40 to 50 wt % Mn): The name psilomelane is used for a hydrous Mn oxide that contains about 40 to 50 wt % Mn, up to 14 wt % barium and variable amounts of potassium, silicon and water. The mineral occurs in the Wabush deposits as microcrystalline grains, and is present as masses and intergrowths with hematite, pyrolusite, wad and quartz. In a few places colloform psilomelane and associated secondary quartz have been deposited in voids and open spaces in fracture zones.

8.2.2.6.3. Wad (Mn oxide with 18 to 22 wt % Mn): The name wad is used for an earthy mixture of Mn oxides and hydroxides. The mixture contains about 18 to 22 wt % Mn. It is intergrown with psilomelane (Figure 8.8).

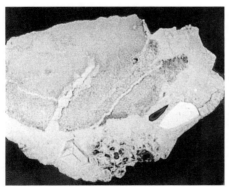

Figure 8.8. SEM photomicrograph of a polished section of psilomelane (light grey) and wad (grey). The psilomelane contains an inclusions of hematite (very light grey). Wabush deposit.

8.2.2.6.4. Ankerite and Siderite: A summary of microprobe analyses of 16 ankerite grains, 4 Mn-siderite grains, and 3 siderite grains from the Carol Lake ores is shown in Table 8.1.

Table 8.1
Analyses of carbonates from Carol Lake ores

Mineral	Fe (wt %)	Mn (wt %)	Mg (wt %)	Ca (wt %)
Ankerite	6 to 15	2 to 7	3 to 6	20 to 24
Mn - siderite	24 to 36	10 to 13	4 to 7	1
Siderite	35 to 39	3 to 4	3 to 5	1

8.2.3. Mining

The ores of the Carol Lake, Wabush and Mount Wright deposits are mined by open pit. Assays of drill cores are used to outline the orebodies, and analyses of blast hole cuttings are used as grade control for mineral processing. The correlation between the predicted recovery of iron minerals using the blast hole data and the actual recovery obtained by the concentrator is sometimes poor. The poor correlations are due to sampling techniques, mineral liberations and other unknown mineralogical and ore characteristics that define the ore variables. Since all the variables are not known, it is difficult to develop a reliable mathematical model; nevertheless, the operators use empirical working models.

8.2.4. Mineral Processing

Spirals, and to a small extent magnetic separators, cones and other gravitational techniques are used to recover the Fe-bearing minerals at the IOC, Wabush and the Quebec Cartier Mining Company (Mount Wright) mineral processing operations. The Mount Wright deposits have the lowest grade ores, but the highest Fe recoveries are obtained from them (Table 8.2). The high recoveries are obtained because the hematite and magnetite grains in the Mount Wright ores are in the appropriate size range for efficient recovery by spirals. In particular:
- Recoveries of liberated hematite and magnetite (Fe oxides) by spirals are high for particles that range from 850 to 150 μm, drop for particles smaller than 100 μm, and are only about 20 % for particles smaller than 37 μm in diameter (Figure 8.9).
- Most of the hematite and magnetite (Fe oxides) in the Mount Wright deposits are coarse-grained and are between 850 and 100 μm in diameter. In contrast, the Fe oxide minerals in the Carol Lake (IOC) and Wabush deposits are about 2 to 2 ½ Tyler sizes finer-grained than the Mount Wright Fe oxides, and are too fine-grained for high recoveries by spirals (Figure 8.9).

Table 8.2.
Grades and Fe recoveries obtained during the 1980's

Mining operation	Ore grade (wt % Fe)	Fe Recoveries (%)	Liberation (Fe oxides) (%)
Mount Wright	32	85	96
IOC	42	68	90
Wabush	35	66	90

8.2.4.1. A case history: determining mineral characteristics that affect mineral processing of IOC ores

The IOC engineers and geologists have continuously investigated techniques that might improve iron recoveries. On-going mineralogical support has been provided by a mineralogist who used optical microscopy, Davis tube analysis, heavy liquid separations, Frantz magnetic separations and point counting to determine the mineral and ore properties. The author studied the ore and process products from 1985 to 1991 using SEM, microprobe, image analysis and materials balancing techniques to determine whether other mineral properties could be detected and used to improve Fe recoveries. The first phase involved determining the mineral characteristics of the ore and concentrator products (Petruk, 1985; Petruk and Pinard, 1988; Petruk, 1990d; Petruk, 1991).

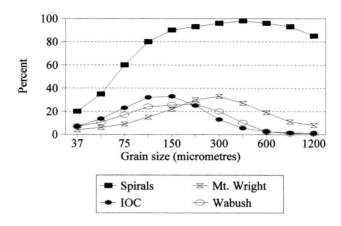

Figure 8.9. Diagram showing recoveries obtained by spirals for different sized particles of Fe oxides (top curve), and size distributions of Fe oxide grains in the Mount Wright, IOC and Wabush deposits.

8.2.4.1.1. Mineral characteristics of ore and concentrator products: During the period of the investigation the IOC concentrator used a dry and a wet grinding circuit to grind the ore (Figure 8.10). One bank of spirals concentrated ore from the dry grinding circuit, and another bank concentrated ore from the wet grinding circuit. In addition cones were used to recover iron minerals from secondary cyclone underflows, and a low intensity magnetic separation circuit (LIMS) recovered magnetite from secondary cyclone overflows and from reground spiral and cone tails.

Samples were collected by IOC personnel from selected points in the concentrator and studied by the author. The samples were sieved into +850, 850-600, 600-425, 425-300, 300-150, 150-75, 75-37 and -37 μm fractions. The sieved fractions were split into sub-fractions. One sub-fraction was analyzed chemically for Fe, and a polished section was prepared from the other sub-fraction. The polished sections were analyzed with the image analyzer to identify the minerals and to determine mineral quantities and mineral liberations. The image analyzer used a microprobe as an imaging instrument, and was set to detect gangue, siderite, goethite, hematite and magnetite in the backscattered electron (BSE) image. The gangue consisted mainly of quartz, but contained some calcite and dolomite and minor amounts of amphiboles, pyroxenes, chlorite, biotite and apatite. The hematite was differentiated from magnetite in the BSE image by using a high electron beam current (20 kV and 30 nA) because the average atomic numbers of these minerals are very close (see Table 2.1). Since the spirals recover both hematite and magnetite as one product, an additional image was produced with the image analyzer and referred to as Fe oxides. The Fe oxides image was produced by either combining the images of hematite and magnetite, or setting an additional window that detected both hematite and magnetite as the same phase.

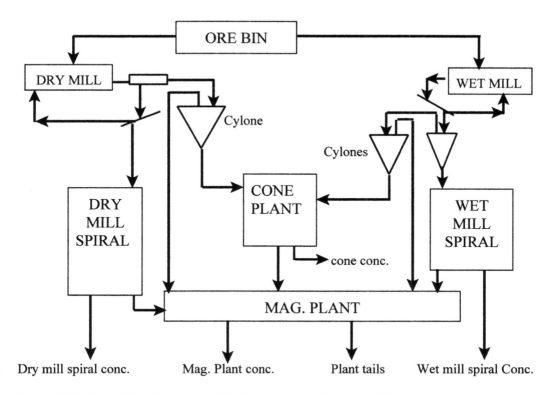

Figure 8.10. Simplified flowsheet of IOC concentrator, March, 1988.

A good correlation was observed between assayed Fe contents and calculated Fe contents from the image analysis data for sub-fractions of concentrates and tails, but a poor correlation was observed when the sub-fractions contained significant amounts of both Fe oxides (hematite and magnetite) and gangue. The poor correlation occurred because the mineral quantities exposed on the surfaces of the polished sections were not representative of mineral quantities in the sub-fractions due to mineral segregation during preparation of the polished sections. The hematite and magnetite, which have specific gravities (S.G.) around 5.2, settled to the bottom of the sample mount (e. g. the face of polished section), whereas the gangue, mainly quartz (S.G. of 2.65), did not settle to the same extent in the araldite mounting medium. Hence the proportion of hematite and magnetite exposed on the surface of the polished section was about 30 % higher than the true proportion. The value of + 30 % varied considerably, depending upon the depth to which the sample was ground during polishing. The mineral quantities were therefore calculated for all samples by using the ratios of hematite:magnetite:goethite:siderite determined by image analysis, and the assayed Fe content to account for the Fe minerals (disregarding Fe in gangue). The gangue content was calculated by difference. An example of mineral quantities in sieved fractions is included in Table 8.3 for IOC Sample 91.018 - wet spiral feed - March 16. 1991. It was observed for all samples studied that the proportion of magnetite to hematite increased with increasing sieve size, regardless of the magnetite content in the samples (Figure 8.11).

Table 8.3.
Mineral contents in sieved fractions (wt %)

Mineral	Size range (µm) of sieved fractions								
	-37	37-75	75-150	150-300	300-425	425-600	600-850	850	Total sample
Gangue	42.6	25.9	35.6	42	43.3	40.2	45.3	51	37.8
Hematite	38.8	56.8	45.6	33.2	27.2	23	23.9	25.2	40.1
Magnetite	15.8	16.2	18	23.8	26.8	34.7	29.3	21.6	20.8
Siderite	1.1	0.7	0.4	0.7	1.7	1.6	0.3	0.6	0.7
Goethite	1.7	0.4	0.4	0.3	1	0.5	1.2	1.6	0.6
Fe (wt %)	40.1	52	45.3	40.9	39.7	42.2	38.7	34.5	43.7
Mag./hem.	0.43	0.3	0.42	0.8	0.99	1.51	1.23	0.86	0.53
Sieve analysis (%)	6.1	15.3	34.5	27.2	7	3.7	2.3	3.9	100

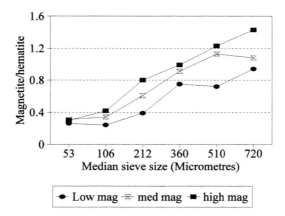

Figure 8.11. Distribution of magnetite/ hematite among different sized sieved fractions. The average magnetite contents in the samples were, in wt %, low mag = 14.5, med mag = 18.5 and high mag = 20.8.

The liberation characteristics of the Fe oxides were determined by image analysis by classifying the particles in incremental steps of 10 % from 0.1 to 100 % Fe oxides in the particles. Table 8.4 is an example of liberation data for Fe oxides in sieved fractions of IOC Sample 91.018 - wet spiral feed - March 16. 1991. It is noted that most of the Fe oxides in all sieved fractions are in particles that contain more than 90 % Fe oxides (e.g. 90 - 99.9 plus 100 % particle categories). This is particularly significant because only particles that contain >90 % Fe oxides should be recovered when producing high grade concentrates. It was also determined that high recoveries were obtained for particles that contained >90 % Fe oxides, and that the recoveries dropped for particles that contained <90 % Fe oxides (Petruk and Pinard, 1988). The term *liberated >90* is, therefore, used in this chapter for particles that contain >90 % of a specific mineral (e.g. Fe oxides, hematite, etc.).

Table 8.4
Liberation data (% in each particle category of each sieved fraction)

Particle category (%)	- 37 μm	37 - 75 μm	75 - 150 μm	150- 300 μm	300 - 425 μm	425 - 600 μm	600 - 850 μm	Total sample
100	73.9	65.7	66.8	47.6	37.7	36.5	16.8	48.8
90 - 99.9	21	25.5	25.5	39.7	43.9	44.1	47	36.7
80 - 89.9	1.9	2.9	2.7	4.2	6.4	5.2	10.5	7.4
70 - 79.9	1.3	1.6	1.4	1.9	3.5	4.2	4.7	1.8
60 - 69.9	0.3	1.2	1.4	1.3	1.5	1.9	3.5	1.3
50 - 59.9	0.5	1.3	0.5	1	0.6	2	3.1	0.9
40 - 49.9	0.2	0.8	0.5	1.2	1.7	1.4	5.9	0.9
30 - 39.9	0.4	0.4	0.3	0.9	1	1.4	3.5	0.6
20 - 29.9	0.3	0.3	0.2	0.9	1.1	1.6	3.4	0.6
10 - 19.9	0.1	0.2	0.3	0.6	1.7	0.7	1.4	0.5
0.1 - 9.9	0.1	0.1	0.4	0.7	0.9	1	0.2	0.5
Total	100	100	100	100	100	100	100	100

The performances of the dry spiral, wet spiral, cone and LIMS circuits, and of the concentrator were evaluated by analyzing a suite of 25 samples that was collected throughout the plant on March 4, 1988. Each sample and sieved fraction was analyzed for quantities of Fe, Fe oxides, hematite, magnetite, silicates and quartz, and liberations of Fe oxides. Materials balance calculations, using MATBAL (Laguitton, 1985), were performed on the data to determine the weight % and recoveries of each phase throughout the flowsheet. Table 8.5 gives the weight % (weight yields), Fe assays and quantities of magnetite, silicates and quartz in the concentrates from each circuit and from the concentrator. The table shows that the concentrator recovered 47 wt % of the concentrator feed at a grade of 66.2 wt % Fe and contained 5.3 wt % silicates. The table also shows that the two spiral circuits recovered 43.3 wt % of the concentrator feed; the cones recovered 1.3 wt % and the LIMS 2.4 wt %.

Table 8.5
Weights and quantities of Fe and minerals in different concentrates and concentrator tails

Products	Weight %	Fe (%)	Magnetite (%)	Silicates* (%)	Quartz (%)
Concentrator feed	100	42.4	16.6	37.8	22.9
Dry grind spirals concentrate	14.5	66.5	24.9	4.4	3
Wet grind spirals concentrate	28.8	65.9	28.6	5.9	3.3
Cone concentrate	1.3	66.5	18.1	5.1	4.6
LIMS concentrate	2.4	67.5	94.6	2.9	2.3
Concentrator concentrate	47	66.2	30.8	5.3	3.2
Concentrator tails	53	21.1	3.6	67.1	

* silicates = mainly quartz and carbonates, but includes traces of chlorite, biotite and amphiboles.

The recoveries by dry spiral, wet spiral, cone and LIMS circuits, and by the concentrator, were evaluated by analyzing the recoveries of *liberated* >*90* Fe oxides to avoid the masking effects of unliberated Fe oxides in silicate-bearing particles. The analysis involved determining, by materials balance calculations, the weight units of *liberated* >*90* Fe oxides in each sieved fraction of the feed to each circuit, and of the concentrate from each circuit (Table 8.6). weight units are defined as *'wt % of respective products with respect to weight of feed'*. The results are plotted in Figures 8.12 to 8.16 as weight units of *liberated* >*90* Fe oxides at each sieve size in the feed and concentrate for each circuit. The mid-size of each sieved fraction is plotted as the sieve size. The top curve in each diagram represents the distribution of *liberated* >*90* Fe oxides in the feed to the circuit and the bottom curve the distribution of the *liberated* >*90* Fe oxides in the concentrate that was recovered from the circuit. The space between the two curves represents the amount of *liberated* >*90* Fe oxides lost to the circuit tails. Figures 8.12 to 8.16 show that the main loss of *liberated* >*90* Fe oxides was in particles smaller than 150 μm, the size where spiral efficiency drops. Figure 8.15 shows that the LIMS circuit had a high recovery of *liberated* >*90* Fe oxide particles smaller than 37 μm and a high loss of *liberated* >*90* Fe oxide particles 37 to 212 μm. Microscopical studies show that most of the Fe oxides in LIMS tails are *liberated* >*90* hematite. This shows that Fe recoveries could be improved significantly with a concentrating unit that recovers *liberated* >*90* hematite grains that are smaller than 150 μm.

Table 8.6
Distributions of *liberated* >*90* Fe oxides in circuit and concentrator concentrates (wt units)

concentrating unit	Weight units			Recovery (%)
	Feed	Concentrate	Tails	
Dry grind spirals circuit	18	13.1	4.9	72.8
Wet grind spirals circuit	37.1	26.8	10.3	72.2
Cone circuit	3.2	1.2	2	37.5
LIMS circuit	19.4	2.3	17.1	11.9
Concentrator	60.5	43.4	17.1	71.7

8.2.4.1.2. Pilot plant tests and mineralogical characteristics of products: The company engineers conducted a series of pilot plant tests (Chung, 1992) to determine whether other flowsheets might improve iron recoveries. A taconite flowsheet showed the most promise and was tested extensively (Figure 8.17). The taconite flowsheet was developed for processing taconite ores in Michigan, USA. It consists of LIMS separator at the front-end, and produces a LIMS concentrate and a LIMS tails as feeds to separate spirals. The LIMS concentrate contains magnetite-rich particles, and is processed with MAG spirals. The LIMS tails consists largely of hematite and silicate particles, and is processed with HEM spirals. The pilot plant was run in parallel with the concentrator wet grind spiral circuit (hereafter referred to as IOC *plant spirals*) so that the performance of the pilot plant could be compared to the performance of the *IOC plant spirals*. A splitter box was used to split part of the concentrator feed to the pilot plant. However, the feed to the *IOC plant spirals* was modified somewhat by a primary cyclone in the IOC concentrator. The primary cyclone removed the fines (about 5 wt % of the feed) and directed them to the cones and magnetite plant. The assays and mineral distributions for the *IOC plant spirals* feed were, therefore, slightly different than for the pilot plant feed. Recoveries in the *IOC*

Dry grind spirals

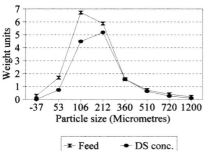

Figure 8.12. Recoveries of *liberated >90* Fe oxides by dry grind spirals.

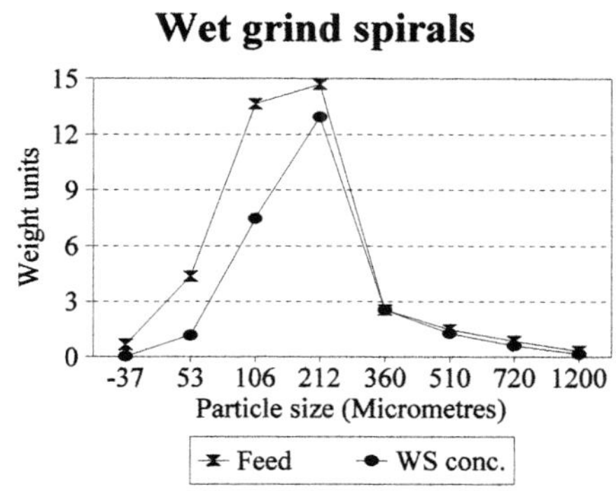

Wet grind spirals

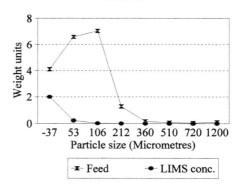

Figure 8.13. Recoveries of *liberated >90* Fe oxides by wet grind spirals.

Cones

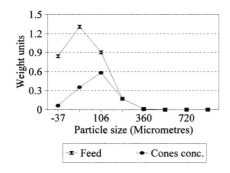

Figure 8.14. Recoveries of *liberated >90* Fe oxides by cones.

LIMS

Figure 8.15. Recoveries of *liberated >90* Fe oxides by LIMS.

Concentrator

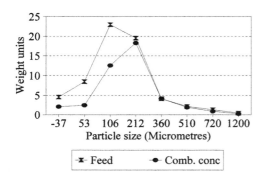

Figure 8.16. Recoveries of *liberated >90* Fe oxides by concentrator.

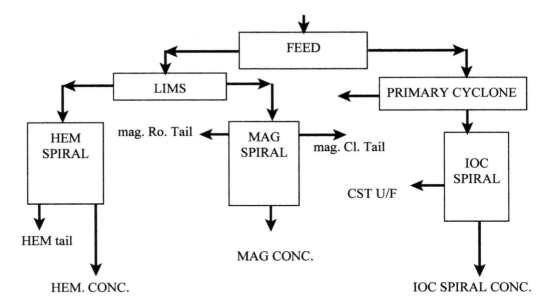

Figure 8.17. Simplified flowsheet of the IOC pilot plant.

plant spirals concentrate were calculated with respect *IOC plant spirals feed*, and recoveries in the pilot plant concentrates were calculated with respect to pilot plant feed. Some of the pilot plant tests produced better recoveries than the *IOC plant spirals*, whereas other tests produced poorer recoveries. A mineralogical study was conducted on a suite of samples from a test when the pilot plant had better recoveries than the *IOC plant spirals* (Petruk et al., 1993). Recoveries of hematite and magnetite by the HEM spirals, MAG spirals and *IOC plant spirals,* are given in Table 8.7.

Table 8.7 shows that a better iron recovery was obtained by the combination of the two spirals in the pilot plant (71.9 %) than by the *IOC plant spirals* (69.6 %). The major difference was due to the behavior of hematite in the HEM spiral circuit, which accounted for about 2/3 of the Fe minerals in the ore. The hematite recovery was 78.8 % by the pilot plant spirals compared to 73.7 % by the *plant spirals*. A comparison of the hematite recovery at each sieve size by the pilot plant HEM spirals and the *plant spirals* is shown in Figures 8.18 and 8.19. The figures give the distributions of the hematite in the spiral feeds and concentrates at each size range in weight units. The area under the concentrate curve represents the recovery, and the space between the two curves represents the loss to spiral tails. The comparison shows that the main improvement by the HEM spirals was a higher recovery of hematite in particles smaller than 150 µm.

The magnetite recovery by the pilot plant (65.8 %) (Table 8.7) was poorer than by the *IOC plant spirals* (71.8 %). A comparison of the magnetite recovery at each sieve size by the pilot plant MAG spirals and the *IOC plant spirals* (Figure 8.20, 8.21) does not show a significant difference. It is interpreted that the higher magnetite recovery by the IOC plant spirals, which process hematite and magnetite together, was due to intimate intergrowths of hematite and recrystallized magnetite in hematite-magnetite particles.

Table 8.7
Assays, distributions and recoveries of Fe, hematite and magnetite in products from pilot plant and *plant spirals*

Product	Fe			Hematite			Magnetite		
	Assay (wt %)	Dist. (%)	Rec. (%)	Analysis (wt %)	Dist. (%)	Rec. (%)	Analysis (wt %)	Dist. (%)	Rec. (%)
Ore	42.6	100		38.4	100		19.1	100	
PPL feed	42.6	100.0		38.4	100.		19.1	100.	
Hs feed	36.5	62.9		46.5	88.9		1.8	7.1	
Hs conc	66.4	**47.9**	**76.1**	91.0	**72.8**	**81.9**	1.7	**2.7**	**38.0**
Ms feed	59.3	37.1		16.0	11.1		66.7	92.9	
Ms conc	69.1	**24.0**	**64.7**	15.7	**6.0**	**54.0**	81.7	**63.1**	**67.9**
PPs conc	67.3	**71.9**	**71.9**	66.7	**78.8**	**78.8**	27.5	**65.8**	**65.8**
Ps feed	44.2	94.5		40.3	95.6		19.5	93.2	
Ps conc	66.7	**65.8**	**69.6**	64.4	**70.5**	**73.7**	30.3	**66.9**	**71.8**

Dist. = distributions; Rec. = recoveries.
Hs feed = HEM spirals feed; Hs conc = HEM spirals concentrate; Ms feed = MAG spirals feed; Ms conc = MAG spirals concentrate; PPL feed = pilot plant LIMS feed; PPs conc = pilot plant spirals concentrate; *Ps* feed = *IOC plant spirals* feed; *Ps* conc = *IOC plant spirals* concentrate.

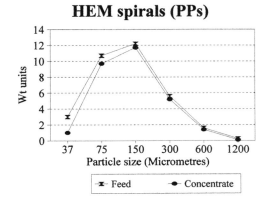

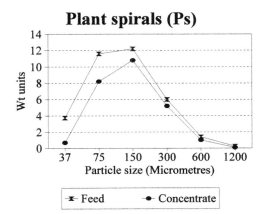

Figure 8.18. Recovery of hematite by pilot plant HEM spirals from feed to HEM spirals.

Figure 8.19. Recovery of hematite by *IOC plant spirals* from feed to IOC *plant spirals*.

MAG spirals (PPs)

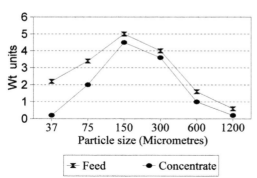

Plant spirals (Ps)

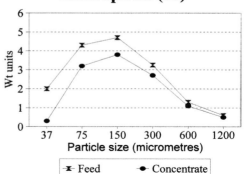

Figure 8.20. Recovery of magnetite by pilot plant MAG spirals from feed to MAG spirals.

Figure 8.21. Recovery of magnetite by *IOC plant spirals* from feed to *IOC plant spirals*.

The reason for the better hematite recovery by the HEM spirals than by the *IOC plant spirals* was investigated by analyzing the feed to the HEM, MAG and *IOC plant* spirals. The mineralogical compositions of the feed, and the liberations, size distributions and textures of the hematite and magnetite were studied. It was determined that the feed to the HEM spirals consisted mainly of hematite and quartz, whereas the feed to the *plant spirals* consisted of hematite, quartz and liberated and unliberated magnetite. This suggests that the presence of some unliberated magnetite may have interfered with the concentration of hematite by spirals.

Liberation analyses (Table 8.8) showed that the *liberation >90* of hematite in the HEM spirals feed was higher than the *liberation >90* of magnetite in MAG spirals feed. The *liberation >90* of magnetite is poorer because some of the magnetite is recrystallized and is intergrown with both quartz and hematite. It is likely that the hematite recovery was better than magnetite recovery because of the higher *liberation >90*.

Table 8.8
Liberation >90 of hematite and magnetite

Product	Hematite (%)		Magnetite (%)	
	HEM spirals	*plant spirals*	MAG spirals	*plant spirals*
Spiral feed	83.9	83.7	75.7	77.5
Spiral concentrate	91.1	86.7	73.6	77.3

A size distribution analysis of the hematite and magnetite particles in the HEM and MAG spiral feeds respectively (Figure 8.22) showed that the size range of hematite particles was narrower than the size range of magnetite particles. As spirals perform better on material that has a narrow size range, part of the reason for the better hematite recovery than magnetite recovery is the narrower size range of the hematite particles.

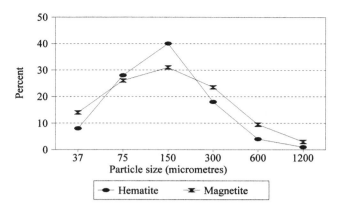

Figure 8.22. Size distributions of magnetite and hematite in MAG and HEM spiral feeds.

The higher liberation of hematite in sieved fractions of the HEM spiral feed than of magnetite in sieved fractions of the MAG spiral feed can also be observed by a qualitative examination of polished sections. In particular, more of the magnetite in the +850 μm fraction of the MAG spiral feed is unliberated than of the hematite in the same sized fraction of the HEM spiral feed (Figures 8.23 and 8.24). Figure 8.25 shows that the magnetite in the -53 μm fraction of the MAG spiral feed is finer-grained than the hematite in the same sized HEM spiral feed (Figure 8.26). Although the magnetite in the -53 μm fraction is finer-grained than the hematite, a lower proportion of the fine-grained magnetite than of fine-grained hematite is liberated.

In summary, the mineralogical investigations have shown that a higher recovery of hematite than of magnetite was obtained by spirals because the hematite had a higher liberation and a narrower size range of hematite particles. Furthermore, a higher recovery of hematite was obtained by the HEM spirals than by the *IOC plant spirals* because the feed to the HEM spirals had a simpler mineral composition.

8.2.4.1.3. Modifications to IOC plant: The pilot plant tests and mineralogical studies showed that it is possible to recover more hematite from the IOC ores with appropriate flowsheet modifications. The mineral processing engineers installed a hematite scavenger spiral circuit, using Reichert spirals (Penney, 1996), in the IOC concentrator to recover the moderately fine-grained *liberated* >90 hematite from the LIMS circuit tails. An additional 3.4 % total iron recovery was obtained. The additional recovery was obtained partly because the LIMS circuit tails have the characteristics of the HEM spiral feed (e.g. simple mineralogical composition), and partly because the LIMS circuit had a regrind circuit that reground the wet and dry spiral tails and liberated additional hematite from coarse-grained particles. It is assumed that Reichert spirals recovered the largest hematite particles in the LIMS tails, but not the small ones.

Subsequent pilot plant tests, in parallel with the IOC concentrator, indicated that screening the hematite scavenger spiral circuit feed with Derrick screens using 0.35 mm (350 μm) openings improved iron recoveries an additional 2.5 % (Penney, 1996). These tests showed that removal of the large particles made it easier to optimize the Reichert spirals. As the large particles were not studied it can only be assumed that they were largely barren silicates (quartz).

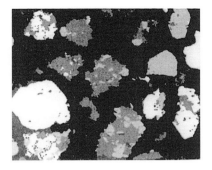

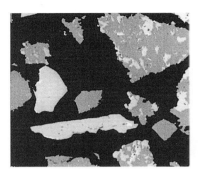

Figure 8.23. SEM photomicrograph of a polished section of the +850 μm fraction of the MAG feed. It shows free magnetite (white), large grains of unliberated magnetite associated with hematite and quartz; and free hematite(grey) as well as small grains of unliberated hematite in quartz (dark grey).

Figure 8.24. SEM photomicrograph of a polished section of the +850 μm fraction of the HEM feed. It shows free hematite (grey), small grains of unliberated hematite in quartz (dark grey), and very small grains of unliberated magnetite (white) in quartz.

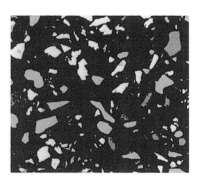

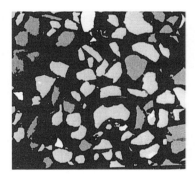

Figure 8.25. SEM photomicrograph of a polished section of -53 μm fraction of MAG feed showing magnetite (light grey) and hematite (grey). The quartz is very dark grey and cannot be seen in the photomicrograph.

Figure 8.26. SEM photomicrograph of a polished section of -53 μm fraction of HEM feed showing hematite (grey), magnetite (white) and quartz (dark grey).

8.2.4.2. Mineralogical characteristics of ore from the Wabush Deposits

A suite of samples from the rougher and cleaner spiral circuits of the Wabush Mines concentrator (Figure 8.27) was studied by L. Lewczuk at New Brunswick Research and Productivity Council, under a Mineral Development Agreement between the Federal Department of Energy, Mines and Resources, Newfoundland Department of Mines, and Wabush Mines Limited, during the period of 1986 to 1988. The objective of the study was to determine the mineralogical characteristics of the ore and process products (Lewczuk, 1988). Subsequently the author studied, at CANMET, a composite sample of concentrator tails for the period of January to June, 1990 (Petruk, 1992), to determine whether more iron can be recovered from the tails.

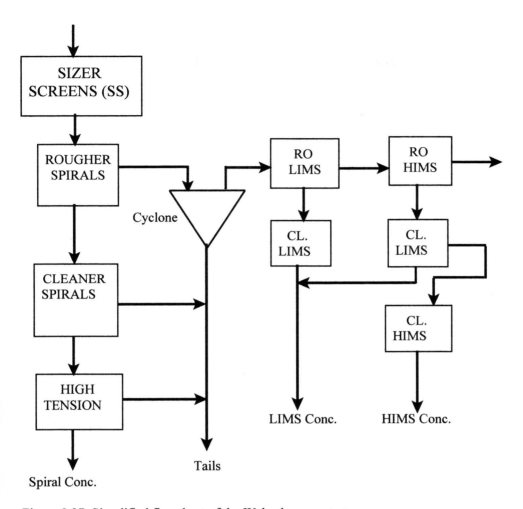

Figure 8.27. Simplified flowsheet of the Wabush concentrator.

The ore from the Wabush deposit contains significant amounts of Mn minerals and goethite (see sections 8.2.1.2, 8.2.2.4, 8.2.2.5 and 8.2.2.6 of this chapter) (Petruk, 1979; Lewczuk, 1988). The assays for Fe, Mn and SiO_2 in samples of the rougher spirals feed (SSU), rougher spirals concentrate (RSC), the cleaner spirals concentrate (CSC) (from Lewczuk, 1988), and of the concentrator tails (CT) (Petruk, 1992) are included in Table 8.9. It is noted that the grade of the cleaner spirals concentrate is low (59.6 wt % Fe), as it contains 7.5 wt % SiO_2 and 1.4 wt % Mn.

The mineralogical study was conducted by treating hematite and magnetite as one phase (referred to as hem-mag), the Mn-bearing minerals (pyrolusite, psilomelane, wad, and Mn goethite) as another phase, Mn-free goethite as a third phase, and gangue (mainly quartz) as a fourth phase. The quantities of the *phases* in the samples are included in Table 8.9.

Table 8.9
Assays and *phase* contents in samples and wt % represented by samples

Product	Weight % of product	Assays (wt %)			*Phase* contents (wt %)			
		Fe	Mn	SiO$_2$	Hem + mag	Mn minerals	Goethite	Gangue
SSU	100.0	35.3	1.7	43.0	47.9	3.8	4.7	43.6
RSC	52..5	53.3	1.8	14.8	77.6	3.8	2.9	15.7
CSC	43.8	59.6	1.4	7.5	86.4	4.0	2.1	7.5
CT tails		20.8	2.0	63.6	21.8*	3.2	7.3	67.7

SSU = Sizer screens underflow = Rougher spirals feed; RSC = Rougher spirals concentrate;
CSC = Cleaner spirals concentrate; CT tails = Concentrator tails.
* hematite = 20 %, magnetite = 1.8 %

The distributions of elements and *phases* throughout the spiral circuits were calculated by a materials balancing program. The results, given in Table 8.10, show that the cleaner spirals concentrate (CSC) recovered 79 % of the hem-mag and 19.4 % of the goethite, accounting for 74 % of the iron in the spiral circuit feed. The cleaner spirals concentrate also contained 7.6 % of the SiO$_2$, and 45.9 % of the Mn minerals accounting for 36.1 % of the manganese.

Table 8.10
Distributions of elements and minerals (%)

Product	Elements (%)			Minerals (%)			
	Fe	Mn	SiO$_2$	Hem-mag	Mn minerals	Goethite	Gangue
SSU	100.0	100.0	100.0	100.0	100.0	100.0	100.0
RSC	79.3	55.6	18.0	85.1	53.0	31.9	18.9
CSC	74.0	36.1	7.6	79.0	45.9	19.4	7.5

SSU = Sizer screens underflow = Rougher spirals feed; RSC = Rougher spirals concentrate;
CSC = Cleaner spirals concentrate.

Particles containing >85 % of a phase were classified as *liberated >85*. *Liberation >85* data for the hem-mag, Mn minerals, goethite and gangue are given in Table 8.11 as percent of mineral in particles that contain more than 85 % of mineral. The table shows a high *liberation >85* for hem-mag (91.2 %) and gangue (94.5) in the rougher spirals feed and spiral concentrates, which reflects the weak bonding between the hematite and quartz due to the friable nature of the Wabush ore. The *liberation >85* of goethite decreased progressively from the feed to rougher spirals concentrate to cleaner spirals concentrate, which indicates that both spiral circuits rejected *liberated >85* goethite. The *liberation >85* of Mn minerals increased in the cleaner spirals concentrate, which indicates that the Mn minerals were concentrated by spirals because the specific gravity of Mn minerals is close to the specific gravity of hematite and magnetite. It is noted that the *liberation >85* of the gangue (mainly quartz) is high in the cleaner spirals concentrate (85.1 %). Obviously the spirals did not reject *liberated >85* quartz efficiently. The reason for this is not known, but it is possible that the spirals feed contained impurities (e.g. slimes as well as large particles) that interfered with spiral performance.

Table 8.11
Liberations >85 of phases (% in particles with >85 % of phase)

Product	liberation >85 (%)			
	Hem-mag	Mn minerals	Goethite	Gangue
SSU	91.2	46.2	62.8	94.5
RSC	91.1	45.0	40.7	87.3
CSC	91.1	71.8	29.5	85.1
CT tails	77.0	41.8	45.0	91.4

SSU = Sizer screens underflow = Rougher spirals feed;
RSC = Rougher spirals concentrate; CSC = Cleaner spirals concentrate;
CT tails = concentrator tails

The distributions, in weight units, of hem-mag, goethite and Mn minerals in the spiral feed (SSU) and in the cleaner spirals concentrate (CSC) by particle size are given in Figures 8.28, 8.29 and 8.30. The top curve in each figure shows the amount of each *phase* in the feed, and the bottom curve shows the amount in the concentrate. The area under the bottom curve represents the recovery and the space between the two curves represents the loss. The figures show that:
- High recoveries were obtained for hem-mag in particles larger than 150 μm, but recoveries decreased for smaller particles (Figure 8.28).
- Most of the large Mn minerals (+212 μm), and a significant proportion of the smaller ones were recovered in the cleaner spirals concentrate (Figure 8.29).
- Goethite recovery was low at all sizes (Figure 8.30).

Details of recoveries of hem-mag, Mn minerals and goethite are shown in Figure 8.31 by bar graphs in a series of diagrams. The diagrams compare the quantities of *liberated >85* (free) and unliberated phases in various products as follows:
- The behavior of hem-mag is determined by comparing the diagrams for the cleaner spirals concentrate and tails. The diagrams show that most of the *liberated >85* (free) hem-mag recovered in the cleaner spirals concentrate was in particles larger than 82 μm, whereas most of the hem-mag in the cleaner spirals tails was in particles smaller than 150 μm(largely *liberated >85*). This shows that the spirals recovered coarse-grained *liberated >85* particles.
- The Mn minerals are compared by analyzing the rougher spirals feed and cleaner spirals concentrate. The Mn minerals in the cleaner spirals concentrate have a much higher >85 *liberation* (free) for particles 82 and 425 μm than the rougher spirals feed. This indicates a concentration of *liberated >85* Mn minerals by the spirals.
- The goethite is compared by analyzing the rougher spirals feed and cleaner spiral concentrate. The amount of *liberated >85* (free) goethite in the cleaner spirals concentrate is low indicating that *liberated >85* goethite was rejected by the spirals.

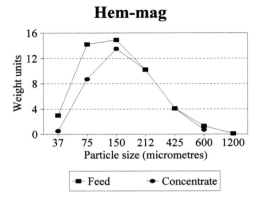

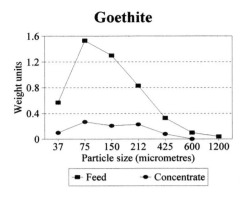

Figure 8.28. Distributions of hem-mag in rougher spirals feed (top curve) and in cleaners spirals concentrate (bottom curve) related to particle sizes. The area under the bottom curve represents the recovery in the cleaner spirals concentrate.

Figure 8.29. Distributions of goethite in rougher spirals feed (top curve) and in cleaners spirals concentrate (bottom curve) related to particle sizes. The area under the bottom curve represents the recovery in the cleaner spirals concentrate.

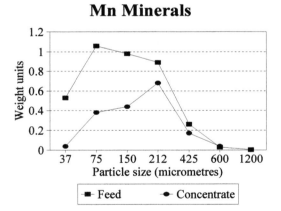

Figure 8.30. Distributions of Mn minerals in rougher spirals feed (top curve) and in cleaners spirals concentrate (bottom curve) related to particle sizes. The area under the bottom curve represents the recovery in the cleaner spirals concentrate.

Most of the silicates (mainly quartz) in the cleaners spirals concentrate are largely liberated (free), and most particles are between 75 and 300 µm in size. The company uses a high tension separator (electrostatic separator) to remove the silicates from the cleaner spirals concentrate.

The concentrator tailings sample was studied to determine whether some of the hem-mag was recoverable. It was determined that 77 % of the hem-mag was *liberated >90* (e.g in particles that contain >90 % hem-mag). It was also determined that the unliberated hematite was associated with quartz and goethite, and that 82 % of the goethite in the sample was associated with unliberated hematite.

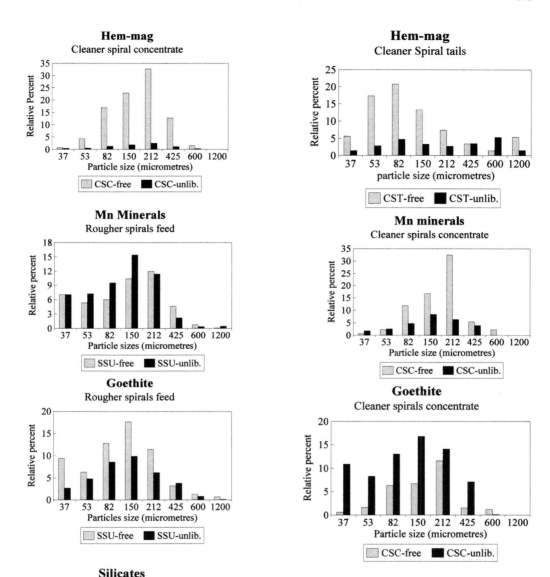

Figure 8.31. Bar diagrams showing the recoveries of *liberated >85* and unliberated grains of hem-mag, Mn minerals and goethite at different size ranges by comparing the amounts in feeds and concentrates, or in concentrates and tails. The relative proportions of *liberated >85* silicates in the cleaner spirals concentrate are shown in the left diagram at the bottom of the page.

The distributions of total hem-mag and of *liberated >90* hem-mag, in particles of different sizes (Figure 8.32), show that much of the hem-mag is in particles that are finer-grained than 150 μm, which is too fine-grained for good recovery by spirals. It may be possible to recover the fine-grained *liberated >90* hem-mag with appropriate equipment (e.g. HIMS).

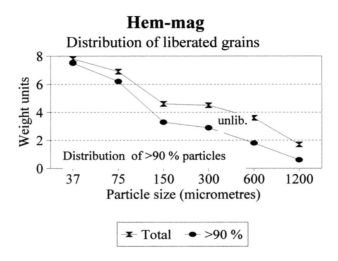

Figure 8.32. Distributions of hem-mag in plant tails. The top curve represents the total amount in all particles related to particle size, and bottom curve represents *liberated >90* hem-mag. The area under bottom curve represents distribution of *liberated >90* particles.

8.2.4.3. Mineralogical Characteristics of ore from the Mount Wright deposits and evaluation of QCM concentrator

The Mount Wright deposits are mined and processed by Quebec Cartier Mines Limited (QCM). The major minerals in the ores are hematite and quartz, and the minor to trace minerals are biotite, amphibole, carbonates, goethite, magnetite, apatite, garnet, tourmaline, rutile, ilmenite and pyrite. The ores contain 30 to 32 wt % Fe, and are processed to produce concentrates that contain around 66 wt % Fe. The QCM concentrator consists of a crushing plant, autogenous grinding mills, Robins sizing screens, and rougher, cleaner and recleaner GEC spirals (Wilson et al., 1990) (Figure 8.33). The tails from the cleaner spirals are recirculated through the autogenous mill for further regrind and liberation, and the tails from the recleaner spirals are recirculated through the rougher spirals. The rougher spiral tails are the concentrator tails.

Several suites of samples were collected from the concentrator by company personnel and characterized at CANMET (Wilson et al., 1990; Petruk and Pinard, 1976; Petruk, 1984). One suite of samples was collected when the sizing screen openings, which define the largest size of particles entering the spirals circuits, were set at 2 mm, and another suite when the sizing screen openings were set at 1.6 mm. The recovery of Fe with the 2mm openings was 84.4 %, and with the 1.6 mm openings 85.4 %. The two suites of samples were characterized to determine the reasons for the slightly different recoveries, and to evaluate whether higher recoveries can be obtained (Wilson et al., 1990).

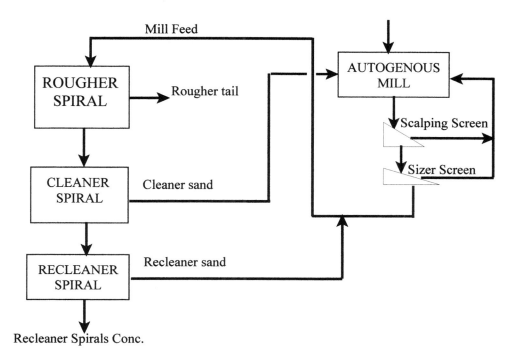

Figure 8.33. Simplified flowsheet of QCM concentrator.

The samples were assayed for Fe and Si and sieved into +850, 600-850, 425-600, 300-425, 212-300, 150-212, 106-150, 75-106, and -75 μm fractions. Polished sections were prepared from each sieved fraction and studied with the image analyzer to determine mineral quantities and liberations of hematite (including minor magnetite and goethite). The mineral quantities were converted to weight percent using specific gravities of the minerals. A BILMAT materials balancing program (Hodouin and Flament, 1985) was used to calculate the flow of solids in tonnes per hour (TPH), and converted to weight units (e.g. in relation to 100 weight units of Fe minerals in the rougher spiral feed). The BILMAT materials balancing program also calculated the distributions of Fe, *liberated >90* hematite, *50 to 90 %* hematite (e.g. particles containing 50 to 90 % hematite), *10 to 50 %* hematite, and *<10 %* hematite throughout the flowsheet.

The distributions of the *liberated >90* hematite on a size by size basis were similar for material produced with the 2 mm and 1.6 mm sizing screen openings. Data for the 1.6 mm screen openings are therfore presented here to describe the behaviour of hematite in the concentrator. The recovery of *liberated >90* hematite in the rougher spiral concentrate from the rougher spiral feed was high for particles of all sizes coarser than 75 μm, and low for particles smaller than 75 μm (Figure 8.34) . In contrast the the cleaner and recleaner spirals lost some of the *liberated >90* hematite particles that were coarser-grained than 300 μm and finer-grained than 75 μm (Figure 8.34, 8.35), but recovered most of the particles 106 to 212 μm in diameter. This shows that the cleaner and recleaner spirals did not recover the large *liberated >90* hematite particles (>300 μm) as efficiently as they recovered medium sized (106 to 212 μm) particles.

Most of the *liberated* >90 hematite in the concentrator tails (Figure 8.34) was finer-grained than 75 μm, and only a small amount was in the coarser size fractions. Hence, the coarse-grained *liberated* >90 hematite rejected by the cleaner and recleaner spirals was not lost to concentrator tails.

The cleaner spiral tails was recirculated through the autogenous mill where the coarse-grained *liberated* >90 hematite was apparently reground. It was subsequently recirculated through the spiral circuits and must have been recovered. Similarly the recleaner spiral tails was recirculated through the spiral circuit and the coarse-grained *liberated* >90 hematite must have been recovered.

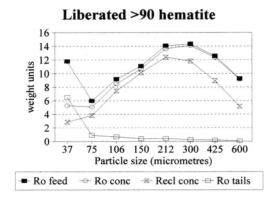

Figure 8.34. Distributions of *liberated* >90 hematite of 1.6 mm sample suite in rougher spirals feed, rougher spirals concentrate, recleaner spirals concentrate and rougher spirals tails (e.g. concentrator tails).

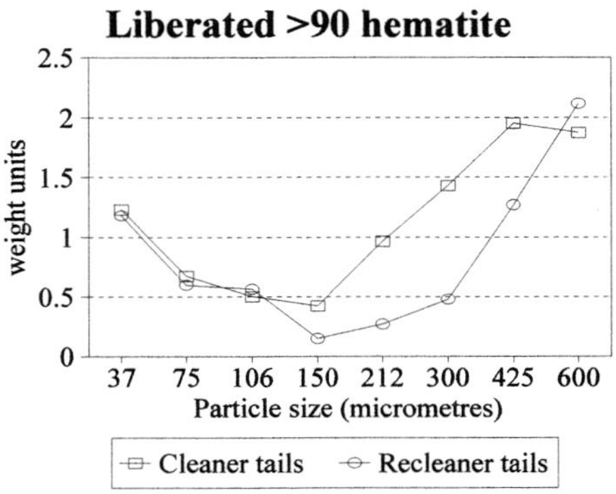

Figure 8.35. Distributions of *liberated* >90 hematite of 1.6 mm sample suite in cleaner spirals tails and recleaner spirals tails.

The liberation of hematite among different types particles produced by using the 2mm and 1.6 mm screens is shown in Table 8.12. The table shows that there was a significant increase in percent of *liberated* >90 hematite with the smaller screen opening. Nevertheless, the *50 to 90* % hematite category (1.6 mm screen opening) still contained nearly enough hematite (3.6 % of hematite in rougher feed) to consider recovering.

Table 8.12
Liberation characteristics of hematite (%) in rougher spirals feed

Particle categories (% hematite)	2 mm opening	1.6 mm opening
liberated >90 hematite	87.9	94.1
50-90 % hematite	8.3	3.6
10-50 % hematite	2.9	1.7
<10 % hematite	0.9	0.6
Total	100.0	100.0

Much of the *50 to 90 %* hematite in the rougher spirals feed produced with the 2mm opening screen was in particles that are larger than 212 μm (Figure 8. 36). In contrast the quantity of large particles in the rougher spirals feed produced with the 1.6 mm opening screen was less. Similarly the quantity of large *50 to 90 %* hematite particles in the cleaner spirals tails obtained from material produced with the 2mm opening screens was higher than from material produced with the 1.6 mm opening screen (Figure 8. 37). The total quantity of unliberated hematite in rougher spirals tails (e.g. concentrator tails) was slightly higher for the 2mm opening screens than for the 1.6 opening screens (2.6 wt units VS 2.1 wt units) (Figure 8. 38). These results suggest that, since the liberation of hematite from *50 to 90 %* hematite particles increased when the 1.6 mm sizing screen openings were used, the increased liberation accounted for the increased Fe recovery. The Liberations characteristics, described above for the rougher spirals feed and the cleaner spirals tails, indicate that the 1.6 mm sizer screen opening is superior to the 2.0 mm sizer screen opening for the QCM ore.

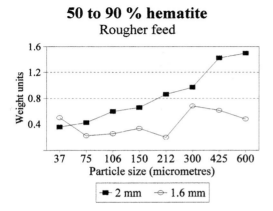

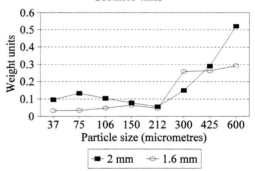

Figure 8.36. Distributions of *50 to 90 %* hematite in rougher spirals feed ground to minus 2 mm and 1.6 mm, respectively.

Figure 8.37. Distributions of *50 to 90 %* hematite in cleaner spirals tails obtained from material that was ground to minus 2 mm and 1.6 mm, respectively.

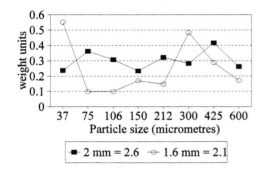

Figure 8.38. Distributions of unliberated hematite in rougher spirals tails obtained from material that was ground to minus 2 mm and 1.6 mm, respectively.

A comparison of the size distributions of *liberated >90* hematite in the rougher spirals feed produced with the 2 mm and 1.6 mm screens shows that *liberated >90* hematite produced with the 1.6 mm sizer screen opening was detectably finer-grained than *liberated >90* hematite produced with a 2 mm sizer screen opening (Figure 8.39).

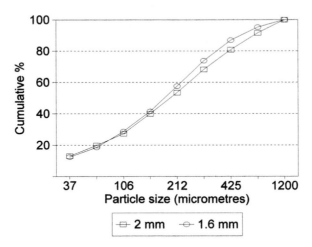

Figure 8.39. Size distributions of *liberated >90* hematite produced with 2 mm and 1.6 mm sizer screen openings of the Robins screens in the grinding circuit. Determined by image analysis.

The mineral characterization studies suggest that QCM concentrator recovers most of the iron that can be recovered with spirals. The possibility for a significant increase in Fe recovery is to recover the -75 μm *liberated >90* hematite from the concentrator tails. This would involve installing an additional concentrating unit, such as a high intensity magnetic separator (HIMS) or a flotation circuit. The HIMS should operate satisfactorily on the Mount Wright ores because the ores do not contain much magnetite.

8.3. MINETTE-TYPE O0LITIC IRONSTONE, PEACE RIVER IRON DEPOSITS

A mineralogical study was conducted on ironstone from the Peace River iron deposits (Petruk, 1977; Petruk et al., 1977a; Petruk et al., 1977b) as part of a research project on recovering iron from the material. A close correlation was obtained between predictions from the mineralogical study and results from mineral processing laboratory tests (Petruk et al., 1977a). The mineralogical characteristics of the Peace River iron deposit in relation to mineral processing are presented in this chapter.

8.3.1. Mineralogical characteristics of the Peace River iron deposits

The Peace River iron deposits consist of flat-lying bodies (Green and Melon, 1962) of Minette-type (Gross, 1965) oolitic ironstone of late Cretaceous age. The bodies are 1 to 9 m thick and occur throughout an area about 25 km long and 5 km wide. The ironstone is a brownish, earthy friable material that expels water on exposure to dry atmospheric conditions and reabsorbs it under humid conditions. The material in situ has a loss on ignition (LOI) of about 25 wt % and contains about 33 wt % Fe; after drying overnight the LOI is about 10 wt % and the Fe content is about 37 wt % (table 8.13). The ironstone consists of oolites, siderite and earthy fragments embedded in a matrix of ferruginous opal and clastic material (Figure 8.40). The clastic material consists of illite and *Peace River nontronite** cemented by ferruginous opal. The oolites vary in shape from spheroidal to ellipsoidal, and are about 50 to 1,000 μm in diameter. They consist of concentric layers of goethite, *Peace River nontronite*, and amorphous phosphate in variable quantities around cores. The cores are quartz, massive goethite, amorphous phosphate, magnetite and oolite fragments. An average oolite consists of about 45 wt % goethite, 45 wt % *Peace River nontronite*, 5 wt % quartz and 5 wt % amorphous phosphate, and contains about 45 wt % Fe and 15 wt % SiO_2. Although most oolites are composed largely of concentric layers of goethite and *Peace River nontronite*, some consist largely of goethite, whereas others are composed mainly of *Peace River nontronite*.

Table 8.13
Assays of a sample of wet and dry Peace River ironstone (wt %)

	Fe	SiO_2	Al_2O_3	CaO	MgO	P_2O_5	K_2O	MnO	V	LOI
Wet	33.3	17.8	5.3	1.4	0.9	1.3	NA	0.7	0.2	24.1
Dry	36.7	25.2	5.1	1.7	1.0	1.7	0.5	NA	NA	10.0

NA = not analyzed

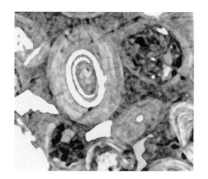

Figure 8.40. Photomicrograph of Peace River ironstone showing an oolite composed of goethite (white) and *Peace River nontronite* (grey) in a matrix of ferruginous opal (dark grey). Fragments of goethite and siderite (undifferentiated) are also present, as well as oolites with cores of clastic material and goethite fragments.

The minerals and oolites have variable compositions; the average mineral compositions, determined by microprobe analyses, are given in Table 8.14. It is noteworthy that both the ferruginous opal and the *Peace River nontronite* contain water. Both absorb water under humid conditions and expel it upon drying. Ferruginous opal expands when wet and tends to make the material firm; upon drying it shrinks and cracks and the material tends to crumble. The *Peace River nontronite* takes water between the crystal structure layers, but does not appear to expand much when wet or to shrink significantly upon drying (Petruk et al., 1977b).

Peace River nontronite is largely an amorphous nontronite containing about 40 wt % of nearly sub-microscopic goethite (Petruk et al., 1977a, 1977b).

Table 8.14
Average partial compositions of minerals in dried material (wt %)

Element or oxide	Goethite (wt %)	Nontronite* (wt %)	Ferruginous opal (wt %)	Siderite (wt %)	Illite (wt %)	Phosphates (wt %)
Fe	49.9	39.0	23.9	46.3	25.5	7.8
SiO_2	6.0	21.0	35.3		34.8	8.1
Al_2O_3	4.6	4.9	5.3	0.9	10.1	3.0
CaO	0.4	0.2	2.1		1.9	42.9
P_2O_5	1.6	0.8	NA		NA	29.2
K_2O	0.3	NA	NA		0.5	1.0
LOI	14.1	19.3	16.9		NA	NA

NA = not analyzed; *Peace River nontronite*

The quantities of oolites and minerals in the Peace River ironstone were determined by image analysis (Table 8.15). The weight percent and distributions of Fe, SiO_2 and P_2O_5 contributed by each mineral were calculated using the mineral quantity data and average mineral compositions reported in Table 8.14. A relatively close correlation was obtained between the calculated element and oxide contents and the assay data (compare highlighted numbers in Table 8.15 with dry assay values in Table 8.13), which indicates that the mineral quantities reported in Table 8.15 are relatively accurate. The results for Fe distributions (Table 8.15) show that only about ½ of the Fe occurs in goethite plus siderite. Most of the remaining Fe occurs as a constituent of the *Peace River nontronite*, and some as a constituent of ferruginous opal. High Fe recoveries are therefore predicated on recovering the *Peace River nontronite* and possibly the amorphous phosphate that occurs in the oolites, as well as the goethite and siderite.

It was calculated that, with perfect separation and recovery of goethite, siderite, *Peace River nontronite*, and amorphous phosphate, a concentrate containing about 43 wt % Fe and 14 wt % SiO_2, with an Fe recovery of about 88 %, could be obtained (Table 8.16). In contrast a perfect separation and recovery of only goethite would yield a concentrate assaying about 49 wt % Fe and 6 wt SiO_2 with an Fe recovery of about 43 %.

As perfect separation and recovery is not possible with any ore, and particularly not with a friable ore that powders readily, an estimate was made of a possible recovery by grinding the ironstone to -1200 µm, screening it into sieved fractions, and determining the quantities of minerals and oolites in each sieved fraction. About 80 % of the goethite, 70 % of the nontronite and 75 % of the siderite were in particles that were large enough to be concentrated by gravitational or magnetic separation techniques (+ 75 µm). These particles accounted for about 65 % of the Fe, which was interpreted as *recoverable Fe* by mineral processing techniques. Assuming a 90 % recovery of the *recoverable iron*, an expected recovery would be around 60 %.

Table 8.15
Analyzed mineral quantities and calculated proportions of Fe, SiO_2 and P_2O_5.

Mineral	Weight % in ironstone	Wt % and distribution					
		Fe		SiO_2		P_2O_5.	
		wt %	Dist.%	wt %	Dist.%	wt %	Dist.%
Goethite	32	16.0	43.8	3.9	15.7	0.54	31.6
Siderite	7	3.2	8.8	--	--	--	--
Peace River nontronite	32	12.6	34.5	6.4	25.8	0.29	16.9
Ferruginous opal	15	3.6	9.9	5.3	21.4	--	--
Quartz	8	--	--	8.0	32.3	--	--
Amorphous phosphate	3	0.2	0.5	0.2	0.8	0.8	51.5
Illite	3	0.9	2.5	1.0	4.0	--	--
Total	100.0	**36.5**	100.0	**24.8**	100.0	**1.71**	100.0
Contribution by oolites (%)	68	29	79	13.6	55	1.5	88

Table 8.16
Grades and recovery/distribution with perfect separation and recovery

Element or oxide	Grade (wt %)	Recovery (%)
Fe	43.2	88
SiO_2	14.2	42
Al_2O_3	4.2	75
CaO	1.6	80
P_2O_5	2.3	100
H_2O	13.6	80

8.3.2. Laboratory mineral processing tests

Laboratory mineral processing tests were conducted by gravitational (tabling), high intensity magnetic separation (HIMS) and flotation techniques (Petruk et al., 1977a). The best recoveries and grades obtained by the three techniques, given in Table 8.17, show that 56.8 to 63.2 % of the Fe was recovered by the three techniques in concentrates grading 42.1 to 43.3 wt % Fe.

Table 8.17
Grades and recoveries of concentrates obtained by laboratory mineral processing tests.

	Weight %	Grade (wt %)		Recovery (Fe %)
		Fe	SiO$_2$	
Tabling	52.5	42.6	17.8	59.0
Tabling*	4.4	48.0	5.1	5.6
HIMS	50.7	42.1	15.4	56.8
Flotation	54.2	43.3	13.3	63.2

* goethite concentrate

Reduction roasting tests were conducted to determine whether higher Fe recoveries could be obtained by converting the goethite, siderite and the nearly sub-microscopic goethite in the *Peace River nontronite* to magnetite. The roasted material was then processed by Low intensity and high intensity magnetic separations (LIMS and HIMS) to recover the magnetite-bearing particles. A concentrate assaying 52.3 wt % Fe and 13.3 wt % SiO$_2$ was obtained with an Fe recovery of 80.3 %. The mineralogical characteristics of the roasted products were not studied. Other pyrometallurgical tests have been conducted on the Peace River material by Alberta Research Council and private companies, but the results are not available.

8.4. SUMMARY

8.4.1. Carol Lake ores

- The Carol Lake ores contain about 34 to 45 wt % Fe.
- The main Fe-bearing minerals are hematite and magnetite, trace Fe-bearing minerals are goethite and siderite, and the main gangue mineral is quartz.
- The quartz has leached along grain boundaries and separates readily from hematite and granular magnetite during grinding.
- Some of the magnetite has recrystallized and is intimately intergrown with hematite and quartz.
- The magnetite:hematite content is variable, but the average magnetite:hematite ratio is about 1:2.
- The ore types range from a low magnetite ore (average about 14 wt % magnetite) to a high magnetite ore (average about 21 wt % magnetite). The hematite content varies from about 35 to 47 wt %. In sieved fractions of ground ores, the magnetite:hematite ratio increases with increasing sieve size, ranging from about 0.3 for 37.5 -75 µm fractions of all ore types, to 1.0 for 600 to 850 µm fractions of low magnetite ores, to 1.4 for 600 to 850 µm fractions of high magnetite ores.
- The concentrator is designed and operated to recover particles that contain more than 90 % Fe-oxides (>90 %) and that are in the proper size range for recovery by spirals.
- By considering the hematite and magnetite as one phase (hem-mag), about 90 % of the hem-mag in the spiral feed was in >90 % hem-mag particles. The remaining hem-mag was in

particles that contain >10 % gangue.
- Only about 65 to 70 % of the hem-mag in the spiral feed was recovered by the spirals, as many of the >90 % hem-mag particles were too small to be recovered efficiently by spirals.
- About 4 % more of the hem-mag (mainly magnetite) was recovered in the magnetite plant after regrinding (e.g. about 4+ % of the hem-mag in the spiral feed was in magnetite-bearing particles rejected by the spirals).
- Most of the hem-mag rejected by the spirals was fine-grained liberated hematite.
- The hematite grains have a narrower size range than the magnetite grains and are liberated more readily than magnetite.
- Many of the large magnetite grains are intimately intergrown with quartz, and are not easily liberated by grinding.
- A higher proportion of magnetite than of hematite occurs as small grains, mainly as inclusions in quartz and hematite.
- Separating the hematite from magnetite with a LIMS, and producing a hematite-quartz stream, makes it possible to optimise the spiral performance and to recover hematite more efficiently.
- The reground magnetite plant tails, which consist of quartz and fine-grained hematite, can be processed to recover the fine-grained hematite. The company has installed a tails circuits composed of screens to remove particles that are larger than 350 μm, and spirals to recover the liberated fine-grained hematite. The tails circuit recovers a significant proportion of the hematite, but a large proportion of the fine-grained hematite is still in the magnetite plant tails. Recovery of the fine-grained hematite will require either HIMS or flotation.

8.4.2. Wabush ore

- An average ore contains about 35 wt % Fe, 1.8 wt % Mn, and less magnetite than the Carol Lake ores.
- The quartz has been extensively leached along grain boundaries leaving a minimum of intergranular cohesion.
- The deposits have been extensively oxidized. The oxidation altered the Fe carbonates to goethite, converted some of the magnetite to martite, and released the Mn from the carbonates and some of the magnetite.
- The manganese was reprecipitated as a variety of Mn oxides including pyrolusite, psilomelane, wad, Mn goethite and a constituent of martite.
- Some of the goethite was formed by oxidation of hematite. This goethite does not contain manganese.
- High recoveries of hematite and magnetite particles larger than 150 μm were obtained by spirals, but the recoveries decreased for smaller particles.
- Most of the large Mn minerals (+212 μm), and a significant proportion of the smaller ones were recovered in the cleaner spirals concentrate.
- The recovery of Mn-free goethite was low for particles of all sizes, as goethite tended to be rejected to tails.
- The cleaner spirals concentrate contained a large amount of liberated quartz. The quartz was removed by high tension electrostatic separation.

8.4.3. Mount Wright ores

- The Mount Wright ores contain about 30 to 32 wt % Fe.
- Most of the iron is present as hematite, with only a small proportion as magnetite.
- The quartz has been extensively leached along grain boundaries, and separates readily from hematite during grinding.
- The Mount Wright hematite is coarser-grained than the Carol Lake and Wabush hematite, due to higher metamorphism, hence higher hematite recoveries are obtained by spirals.
- The Mount Wright spirals recovered hematite particles that are coarser-grained than 75 μm.
- The highest hematite recovery by the cleaner and recleaner spirals was for particles 106 to 212 μm in size.
- A sizer screen opening of 1.6 mm for the spirals feed produced a better hematite recovery than a 2 mm opening, because better hematite liberation was obtained with the smaller opening.
- It is possible to recover more Fe by recovering the -75 μm *liberated >90* hematite from the concentrator tails. This would involve installing an additional concentrating unit, such as a high intensity magnetic separator (HIMS) or a flotation circuit.
- The HIMS should operate satisfactorily on the Mount Wright ores because the ores do not contain much magnetite.

8.4.4. Minette-type Peace River ironstone.

- The Peace River ironstone in situ contains about 33 wt % Fe with a loss on ignition (LOI) of 25 %.
- Upon exposure to dry atmospheric conditions the ironstone expels water and crumbles readily. The exposed ironstone contains 37 wt % Fe with a LOI of 10 %.
- The ironstone consists of oolites, siderite and earthy fragments embedded in a matrix of ferruginous opal and clastic material.
- The clastic material consists of illite and *Peace River nontronite** cemented by ferruginous opal.
- The oolites vary in shape from spheroidal to ellipsoidal, are about 50 to 1,000 μm in diameter, and consist of concentric layers of goethite, *Peace River nontronite*, and amorphous phosphate in variable quantities around cores.
- The cores are quartz, massive goethite, amorphous phosphate, magnetite and oolite fragments.
- An average oolite consists of about 45 wt % goethite, 45 wt % *Peace River nontronite*, 5 wt % quartz and 5 wt % amorphous phosphate, and contains about 45 wt % Fe and 15 wt % SiO_2.
- *Peace River nontronite* is largely an amorphous nontronite containing about 40 wt % of nearly sub-microscopic goethite.
- Ferruginous opal contains water. It expands when wet and tends to make the material firm; upon drying it shrinks and cracks and the material tends to crumble.
- with perfect separation and recovery of goethite, siderite, *Peace River nontronite*, and amorphous phosphate, a concentrate containing about 43 wt % Fe and 14 wt % SiO_2, with an Fe recovery of about 88 %, could be obtained.
- It was determined by studying ground and screened products that about 80 % of the goethite, 70 % of the nontronite and 75 % of the siderite were in particles that could be concentrated by mineral processing techniques. These particles accounted for about 65 % of the Fe, which was interpreted as *recoverable Fe* by mineral processing.

- Assuming 90 % recovery of the *recoverable iron*, an expected recovery would be around 60 % of the Fe.
- Laboratory mineral processing tests recovered 57 to 63 % of the iron in concentrates grading 42 to 43 wt % Fe and 13 to 17 wt % Si.
- Reduction roasting tests of the ironstone recovered 80 % of the iron in concentrates assaying 52 wt % Fe and 13 wt % Si.

CHAPTER 9

APPLIED MINERALOGY INVESTIGATIONS OF INDUSTRIAL MINERALS

9.1. INTRODUCTION

Mineralogical studies are conducted on industrial minerals* to determine the mineral characteristics that influence exploration and mineral processing, and to determine the desirable and undesirable properties of the minerals. A complete characterization of some industrial mineral products requires both chemical assays and mineralogical data, as the chemical assays alone may be inadequate. This applies particularly to products that contain impurities composed of many of the same elements as the constituents of the industrial minerals.

Exploration procedures used for industrial minerals vary widely because of the diverse occurrences, properties, locations, applications and sale price of the minerals. Some industrial minerals, such as quartz and limestone, are ubiquitous and little or no exploration is necessary; nevertheless, the minerals in different deposits must be characterized to determine whether they are suitable for industrial use. Exploration for other minerals, such as industrial diamonds, often requires special techniques. Diamond exploration is commonly performed by concentrating tracer minerals such as ilmenite, pyroxenes and chromite from till samples, and performing microprobe analyses on the tracer minerals to determine the quantities of minor and trace elements that are indicative of diamondiferous kimberlites.

Applied mineralogy related to mineral processing of industrial minerals is performed by determining the standard ore characteristics such as mineral identities, mineral quantities, grain sizes, textures and mineral liberations. In some instances special sample preparation techniques are needed to bring out the mineral features, and special analytical techniques must be used to characterize some minerals. For example, occupational health and safety agencies have developed standard routines for analyzing asbestos, and have invoked allowable amounts and sizes of asbestiform particles. In other cases, such as analysis for free silica in dusts from various environments, the laboratories need to develop satisfactory analytical techniques.

The role of applied mineralogy with respect to industrial minerals is briefly addressed in this chapter by using examples of studies made on graphite, talc, wollastonite, garnet and quartz, and by summarizing a technique for characterizing airborne dusts in the workplace.

---

* Industrial minerals include amblygonite, andalusite, anhydrite, asbestos, barite, bentonite, brucite, calcite, celestite, chromite, corundum, industrial diamonds, diatomite, dolomite, feldspar, fluorite, garnet, graphite, gypsum, kaolinite, kyanite, lepidolite, lime, limestone, magnesite, mica, nepheline syenite, petalite, phosphates, potash, pumice, pyrophyllite, quartz, salt, sillimanite, spodumene, staurolite, sylvite, talc, vermiculite, volcanic ash, and wollastonite.

9.2. GRAPHITE

Graphite is produced in earth environments where the carbon source materials (e.g. carbonates, hydrocarbons) have been subjected to extreme metamorphism, and have lost their volatile constituents. A graphitic schist may be an early rock type, and rocks with high concentrations of graphite form graphitic ores.

Applied mineralogy investigations are conducted on graphite ores to help mineral processing engineers develop techniques for recovering graphite and for producing saleable grade concentrates. Special procedures need to be used when characterizing ground materials that contain graphite because the mineral has unique properties. The graphite contents of samples can be determined by standard assays for carbon if the ore does not contain carbonate minerals. If the ore contains carbonate minerals, special assay techniques are needed.

When characterizing graphite in crushed and/or ground materials by mineralogical methods it is necessary to use techniques that differentiate graphite from the mounting medium and from associated silicate and metallic minerals. Polished - thin sections of ground particles embedded in resin can be analyzed with an optical microscope by using a combination of transmitted and reflected light to:
- identify the minerals,
- determine mineral quantities by point count analysis,
- determine liberations of graphite by point count analysis.

A preferable approach is to analyze powdered materials mounted in polished sections, by using an image analyzer that is interfaced with a scanning electron microscope (SEM) as the imaging instrument. Unfortunately, the average atomic number of graphite is the same as the average atomic number (~6) of most epoxy resins that are used as a mounting medium for preparing polished sections. Hence, graphite cannot be differentiated from the mounting medium in the backscattered electron (BSE) image produced with a SEM using standard polished sections. However, by using carnauba wax as the mounting medium, the graphite can be differentiated in the BSE image. Carnauba wax has an average atomic number of about 5.4 (Straszheim et al., 1988; Petruk et al., 1992a).

9.2.1. A case history of characterizing a graphite ore

Products from a graphite operation in the Huntsville region in Ontario, Canada were studied in the Process Mineralogy Laboratory, CANMET, Dept. of Energy, Mines and Resources, Ottawa, Canada during the developmental stages of the operation (Petruk et al., 1992a). The results are presented here as an example of applied mineralogy related to mineral processing of a graphite ore. The objective was to determine mineral characteristics that affect the production of high quality, saleable graphite concentrates. The study involved:
- identifying minerals,
- determining mineral quantities,
- determining liberations of graphite in sized mill products,
- determining distributions of liberated and unliberated graphite in sized mill products,
- determining recoveries of liberated and unliberated graphite in sized mill products,
- determining mineral associations of the unliberated graphite in the mill products.

9.2.2. Method of Analysis

The mineral processing engineer had conducted metallurgical tests on the ore and supplied samples of the feed, concentrate and tailings. The feed had been crushed to minus 1600 μm, and all samples were sieved into 1600 - 300 μm, 300 - 212 μm, 212 - 150 μm, and -150 μm fractions. The sieved samples were separated into sink and float sub-fractions using a heavy liquid with a specific gravity of 2.38 to float graphite concentrates. Polished sections were prepared from the sink and float sub-fractions using carnauba wax as the mounting medium. The minerals in the ore were identified by classical mineralogical techniques, which showed that no carbonate minerals were present. The graphite contents in the sieved fractions and in the sink and float sub-fractions were, therefore, determined by chemical analysis for carbon. The quantities of other minerals in the sink and float fractions were determined by quantitative XRD analysis and image analysis using a SEM as the imaging instrument. The mineral quantities in each sieved fraction and in the samples were then calculated from the XRD, image analysis and chemical assay data by using a combination technique.

Liberations and mineral associations of graphite were determined by image analysis. Distributions of the minerals and recoveries of liberated and unliberated graphite were determined by a materials balancing technique.

9.2.3. Results

The ore contained an average of 2.2 wt % graphite, major amounts of quartz, plagioclase and mica, minor amounts of pyrite, amphibole, pyroxene, chlorite and orthoclase, and trace amounts of titanite, monazite, and pyrrhotite. The quantities of major and minor minerals in the concentrate and tailings samples, given in Table 9.1, show that the concentrate contained a significant amount of silicate minerals. The concentrate was considered to be ideal for the investigation because it contained enough liberated and unliberated graphite to evaluate the behaviour of the mineral.

Table 9.1
Mineral quantities in samples (wt %)

Mineral	Concentrate	Tailings
Graphite	65.6	0.5
Quartz	16.8	45.5
Feldspar	9.8	22.5
Mica	2.7	15.1
Chlorite	0.2	0.3
Amphibole &pyroxene	2.0	8.0
Pyrite	2.0	7.0

Table 9.2 shows that the graphite content was the highest in the coarsest-grained sieved fraction (1600 - 300 μm) of the concentrate, and was much lower in the -150 μm sieved fraction. The table also shows that the quantity of quartz + feldspar was much higher in the -150

μm fraction than in the coarser-grained sieved fractions of the concentrate. The quantity of mica was nearly the same in fractions of all sizes of the concentrate.

Table 9.2
Quantities of graphite and impurities in sieved fractions of concentrate and tailings (wt %)

Size Range	Concentrate			Tailings		
	Graphite	Quartz + feldspar	Mica	Graphite	Quartz + feldspar	Mica
1600 -300 μm	92.7	1.9	2.4	0.6	68.8	15.5
300 -212 μm	88.1	5.0	3.6	0.2	60.1	19.0
212 - 150 μm	87.4	7.1	3.5	0.2	53.4	19.5
-150 μm	55.2	36.4	2.4	0.3	60.6	8.0

Liberation analyses were performed by classifying the particles, in incremental steps of 10 %, from 0.1 to 100 % graphite in the particles. The data were combined and reported as distribution of graphite among particles containing 100 % (apparently liberated), 90 - 99.9 %, 80 - 89.9 %, 10 - 79.9 % and 0.1 - 9.9 % graphite. The results (Table 9.3) show that the amount of liberated graphite (100 % particle category) was highest in the 1600 - 300 μm fraction of the concentrate and decreased with decreasing particle size. In contrast the amount of liberated graphite in the tailings was highest in the 1600 - 300 μm and -150 μm fractions. These results indicate that coarse-grained fractions would produce the highest grade concentrate, but the high liberation of coarse-grained graphite in the tailings indicates that there was a loss of apparently liberated coarse-grained graphite.

Table 9.3
Distributions of graphite among particle categories in sieved fractions (%)

Particle categories *	CONCENTRATE				TAILINGS			
	1600 - 300 μm	300 - 212 μm	212 - 150 μm	-150 μm	1600 - 300 μm	300 - 212 μm	212 - 150 μm	-150 μm
100	95.0	74.0	80.0	71.0	41.8	25.3	36.7	68.8
90-99.9	3.1	24.3	18.5	23.6	7.1	3.4	6.7	1.7
80-89.9	1.3	1.1	0.8	2.6	0.6	0.8	1.7	2.4
10-79.9	0.5	0.6	0.6	2.6	30.0	16.3	16.5	14.8
0.1-9.9	0.04	0.03	0.06	0.2	20.5	54.2	38.4	12.3
Total	100	100	100	100	100	100	100	100

* = % graphite in particles

The recovery of liberated and unliberated graphite was determined by a materials balance calculation (Table 9.4). The apparently liberated graphite (graphite particles that appear liberated in polished sections) was classified as liberated and the remainder as unliberated. The major loss to tailings was in the 1600 - 300 µm fraction as both liberated and unliberated graphite (Table 9.4).

Table 9.4
Recovery and loss of liberated and unliberated graphite (%)

Size range	Concentrate		Tailings		Total
	Liberated	Unliberated	Liberated	Unliberated	
1600 -300 µm	24.2	1.3	6.9	9.6	42.0
300 -212 µm	10.2	3.6	0.4	1.0	15.2
212 - 150 µm	9.2	2.3	0.4	0.6	12.5
-150 µm	19.9	8.1	1.6	0.7	30.3
Total	63.5	15.3	9.3	11.9	100.0

Mineral association data were obtained by determining the average mineral contents of particles that contain 90 -99.9 %, 80 - 89.9 %, 10 - 79.9 %, and 0.1 - 9.9 % graphite in each sieved fraction of the concentrate and tailings. The results for the graphite concentrate (Table 9.5) show that the mica content increased as the graphite content decreased in the particles. Hence, the main associated mineral in graphite-bearing particles was mica. There was also an increase in quartz content in particles that contained less than 80 % graphite. Table 9.3 shows, however, that in the graphite concentrate, the proportion of graphite in particles containing less than 80 % graphite was insignificant.

Table 9.5
Mineral associations in graphite concentrate: average mineral contents in graphite-bearing particles (%)

Mineral	Particle categories (% graphite in particles)			
	90 - 99.9	80 - 89.9	10 - 79.9	0.1 - 9.9
Graphite	97.4	85.1	41.6	3.9
Quartz *	0.8	5.3	17.5	26.6
Mica**	1.5	9.0	39.2	67.8
Pyrite	0.3	0.5	0.9	0.9
Other minerals	0.0	0.1	0.8	0.8
Total	100.0	100.0	100.0	100.0

* included minor amounts of feldspar; ** included minor amounts of chlorite.

9.2.4. Interpretation of Results

The results indicate that:
- The coarse-grained fractions of the concentrate had the highest graphite contents, but there was a loss of coarse-grained liberated graphite to the tailings.
- The -150 µm fraction of the concentrate had the lowest graphite content, the poorest

liberation of graphite, and the highest quartz content.

- There was an association between graphite and mica in the unliberated graphite particles in the graphite concentrate, particularly, in the particles with low graphite contents.

9.2.5. Recommendations

The mineral characterization of the ore indicated that it should be ground fine enough($\sim$ -1200 μm) to liberate and recover the coarse-grained graphite, but coarse enough to produce a minimum of -150 μm material. The -150 μm material should be separated and reprocessed separately.

9.2.6. Beneficiation improvement

The operator developed a process that produced concentrates, which graded over 90 % graphite and contained only minor amounts of particles with less than 80% graphite (e.g. particles with high quartz contents). The particles in the concentrates ranged from about 1200 to 150 μm.

9.3. TALC

Talc is used in a wide variety of applications including ceramics, cosmetics, pharmaceuticals, paints, plastics, paper, insecticides, roofing felt, rubber and others (Collings and Andrews, 1990). There is a relatively high demand for the mineral, and various degrees of purity are required for the different uses. The required purity of a talc concentrate ranges from low grade talc, which is used in applications such as a filler in drywall sealing compounds, to medium grade talc, which is free of abrasives and is used in paper and other applications, to high grade talc, which is used in cosmetics and pharmaceuticals. Most uses require fine grained talc without abrasives, and a minimum of asbestiform particles. It is considered noteworthy that trace amounts of talc, which have the appearance of asbestiform particles (Figure 1), have been observed in talc concentrates (Petruk, 1983). The elongated talc particles have the properties of talc so are not likely a health hazard.

Applied mineralogy is used in exploring for talc, in mineral processing, in the production of various grades of talc, and in quality control of cosmetic grade talc.

9.3.1. Applied Mineralogy in exploration

A geochemical exploration technique that was developed for locating buried talc deposits provides an example of using applied mineralogy in exploration for talc (Blount and McHugh, 1986). The technique was tested by analyzing residual soils in districts containing talc orebodies in Pennsylvania, Alabama, Montana and Washington. Samples were collected from different soil horizons and prepared for analysis by splitting, sieving, grinding and treating with HCl to remove the carbonates. The residuum was analyzed by an XRD technique, which had been proposed by Rex (1970), to determine the amount of talc. The technique uses the strong peaks of the major minerals in the sample. In this case the strong peaks of chlorite, talc and quartz were used.

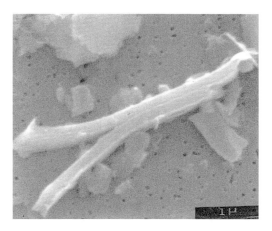

Figure 9.1. Talc in a low grade talc concentrate occurring as an elongated bundle and as irregular grains.

The weight percent talc was determined by the formula:

$$\text{Wt \% talc} = \frac{I_{talc}*K_{talc}}{I_{chl}*K_{chl} + I_{qtz}*K_{qtz} + I_{talc}*K_{talc}} \text{ X } 100$$

I_{qtz}, I_{talc}, and I_{chl} are intensities of the respective XRD peaks for quartz, talc and chlorite. K_{qtz} was set at 1.0, and K_{talc} and K_{chl} were determined by measuring the peak intensities of pure quartz (I_q), talc (I_t) and chlorite (I_c) respectively, and calculating the K_{talc} as I_t/I_q and K_{chl} as I_c/I_q. The geochemical survey pinpointed the locations of talc-bearing deposits.

9.3.2. Applied Mineralogy related to mineral processing

Mineral processing of talc ores involves grinding, flotation and magnetic separation. Mineral processing tests have been conducted at CANMET on many Canadian talc ores, and applied mineralogy studies were conducted on the ores prior to the mineral processing tests, and on mill products during the mineral processing tests. The applied mineralogy studies usually involved identifying the minerals, determining size distributions of the talc, and determining mineral contents by quantitative XRD analysis, point counting or image analysis. For example, the talc ores from Madoc, Ontario, Canada, which were studied in connection with the mineral processing tests, consisted of talc, tremolite, dolomite, calcite, mica, quartz, feldspar and chlorite, and the grind size determined to liberate the talc was 200 μm (Andrews and Soles, 1985; Petruk, 1983). Other minerals found in other talc ores are magnesite, magnetite and trace amounts of antigorite, serpentine, pyrrhotite, pyrite, sphalerite and hematite.

In some instances the quantities of major and minor minerals in mill products from on-going mineral processing tests were routinely determined by quantitative XRD analyses. In other cases the XRD data were used in combination with assays to delineate accurately the talc content in the mill products, because at some mineral combinations the assays for Mg and Ca did not define the purity of the talc sample. The combined analyses were particularly useful when the talc ore contained magnesite, dolomite and tremolite.

As the required purity of a talc concentrate is different for different uses, the grade and purity of the concentrate needs to be defined. The grade is defined by chemical assay and the purity by mineralogical analysis. The mineralogical analysis is performed by standard techniques which involve identifying the minerals, determining mineral contents, and determining the quantity of asbestiform particles by point counting.

9.3.3. Applied mineralogy in quality control of cosmetic products

The impurities allowed in cosmetic talc are present in such small quantities that they cannot be detected by X-ray diffraction. A viable method of analysis has been an optical particle count method such as the one adopted by the food and drug administration in 1973. The procedure, reported by Blount (1990), is:

"weigh out 1 milligram of... talc on each of two microscope slides. Mix the talc with a needle on one slide with a drop of 1.574 refractive index liquid, and then the other with 1.590 liquid, and place on each a ....cover glass sufficiently large so that the liquid will not run out from the edge... and will provide a uniform particle distribution. Fibers counted by this method should meet the following criteria: (i) Length to width ratio of 3 or greater (ii) length of 5 μm or greater (iii) width of 5 μm or less. Count and record the number of asbestos fibers found in each 1 milligram as determined from a scan of both slides with a polarizing microscope at a magnification of approximately 400 X. In the 1.574 refractive index liquid, chrysotile fibers with indices less than 1.574 in both extinction positions may be present; in the 1.590 refractive index liquid, the other five amphibole types of asbestos fibers with indices exceeding 1.590 in both extinction positions may be present. .... A count of not more than 1000 amphibole types of asbestos and not more than 100 chrysotile asbestos fibers per milligram slide constitutes the maximum limit for the presence of these asbestos fibers in talc."

Most users of the method scan 100 fields of view.

A method of separating amphiboles from the cosmetic talc by centrifuging in a heavy liquid was proposed by Blount (1990). The separated amphiboles are removed from the centrifuge tube with a pipette and mounted on a glass slide in a refractive index liquid for mineral identification and point count analysis.

9.4. WOLLASTONITE

New uses and applications of wollastonite have recently created an interest in this industrial mineral. The mineral is used in ceramics, as a filler in paints and plastics, as a substitute for asbestos, and in many other applications (Andrews, 1993). The main property that makes wollastonite a desirable mineral is that it occurs as coarse-grained acicular crystals (aspect ratio 3 - 10), which break into finer-grained acicular crystals during grinding. Furthermore, during cooling of a melt, or by recrystallization, it forms acicular crystals. The addition of acicular crystals of wollastonite to a host mineral contribute a mechanical reinforcement to the resulting material. Some other properties that make wollastonite a desirable industrial mineral are; low thermal expansion coefficient, low sintering temperature (991 - 1196°C), and it fuses readily with alumina and silica. It is used in the ceramic industry because it improves the mechanical properties of the ceramic ware, and greatly reduces warping and cracking of ceramic materials during rapid firing. Its distinctive properties also improve the quality of materials in many other applications.

9.4.1. A case history of characterizing a wollastonite ore

Wollastonite is a contact metamorphic mineral that generally occurs in skarn deposits. A sample of wollastonite-bearing material from the skarn zone in the Little Billy Mine on the northeast shore of Texada Island in British Columbia, Canada was studied with a SEM and image analysis to characterize the material with respect to mineral processing (Lastra et al., 1989). It had been reported that the wollastonite in this deposit is associated with garnet, diopside, magnetite, scheelite and base-metal sulfides (Stevenson, 1945). The minerals found in the sample studied, in wt%, are wollastonite 65, garnet (andradite) 17, quartz plus feldspar 10, dolomite plus calcite 7, and barite plus pyrite, chalcopyrite and magnetite 1. The sample was crushed to minus 1700 μm, and a polished section of the -300 μm fraction was analyzed.

The aspect ratios (maximum diameter/minimum diameter) of liberated wollastonite grains in the -300 μm fractions were determined by measuring the area and maximum diameter of each grain, and calculating average minimum diameter as the area/maximum diameter. The results, given in Table 9.6, show that 70.5% of the liberated wollastonite in the ore sample was in grains whose aspect ratios were >3.

Table 9.6
Distribution of aspect ratios of liberated wollastonite grains (%)

Product	Aspect ratios								
	1 - 3	3 - 4	4 - 5	5 - 6	6 - 7	7 - 8	8 - 9	9 - 10	+10
ore	29.5	21.2	16.4	14.4	6.1	3.0	1.8	3.2	4.4
concentrate	23.3	22.5	16.4	14.4	6.3	4.1	4.9	1.5	6.6

Liberation of wollastonite in the -300 μm particles was measured by determining the distribution of wollastonite among particles containing different amounts of wollastonite, in incremental steps of 10%, from 0.1 to 100% wollastonite, in the particles (Table 9.7) . The results show that 89 % of the wollastonite in the ore was in particles that contain more than 90 % wollastonite. This indicates that high grade wollastonite concentrates can be produced from the ore.

A wollastonite concentrate was produced from material ground to -300 μm by using low intensity magnetic separation to remove the magnetite, high intensity magnetic separation to remove the Fe-bearing silicates, and flotation to remove the carbonates (Andrews, 1993). The concentrate contained, in wt %, wollastonite 88.8, silicates 6.2 and garnet 5.0. About 96 % of the wollastonite in the concentrate was in apparently liberated particles (Table 9.7), and about 76.7 % of the liberated wollastonite was in grains that have aspect ratios >3.0. These results show that a high grade concentrate composed of apparently liberated wollastonite grains can be produced from the ore.

Table 9.7
Liberation of wollastonite

Particle category*	Ore (%)	Concentrate (%)
100.0	43.6	96.0
90 - 99.9	45.2	1.2
80 - 89.9	4.6	0.8
70 - 79.9	2.1	0.7
60 - 69.9	1.2	0.1
50 - 59.9	1.1	0.1
40 - 49.9	0.6	0.6
30 - 39.9	0.4	0.1
20 - 29.9	0.4	0.1
10 - 19.9	0.4	0.2
0.1 - 9.9	0.4	0.1
Total	100.0	100.0

* = % wollastonite in particle.

9.5. GARNET

Garnet is used as an abrasive in a variety of applications (Andrews, 1991). It occurs in metamorphosed rocks, primarily in gneisses and schists. Mineral processing involves concentrating the mineral by gravity, magnetic and flotation techniques. Mineralogical studies are performed to determine the characteristics of the ore and the quality of the garnet. There are no standardized tests for assessing the abrasive quality of garnet, but examinations with an optical microscope, SEM and/or image analysis can indicate whether it has desirable characteristics. Garnet should occur as large, well developed, pure crystals, so that a wide range of size grades, particularly the coarse 1 to 2 mm sizes, would be produced when the ore is crushed. Crystal size and twinning are important measures of abrasive value, since a particle composed of a single crystal is stronger than a particle composed of two or more crystals. The fractures in crushed garnets should be clean, sharp, and angular, as blocky or equidimensional grains produce the best abrasives. In contrast rounded grains have little abrasive value.

The grade of garnet in an ore or concentrate cannot be determined accurately by chemical analysis because of the complex chemical composition of garnet and associated silicates. The most accurate method is a combination of quantitative XRD, image analysis, and a calculation from chemical assays for the elements in the garnet and in the major and minor minerals in the ore. During mineral processing, quantitative XRD, image analysis or point counting may be the most accurate techniques for quality control. On the other hand, a thorough knowledge of the mineralogy of the ore would enable writing a computer program that could be used to calculate the garnet content from chemical assays of samples.

9.5.1. Mineralogical study of a garnet deposit

A mineralogical study of samples from a garnet deposit south of Wabush mines in western Labrador, Newfoundland, Canada is presented here as an example of applied mineralogy related to garnet (Petruk et al., 1992b). The garnet occurs in a garnet-kyanite schist as lumps or nodules of a reddish garnet distributed randomly throughout the rock. Parts of the samples were crushed and analyzed by XRD to identify the major and minor minerals, and to determine mineral contents by quantitative XRD. Other parts of the samples were prepared as polished sections and studied with a SEM to identify the trace minerals and to determine the textures of the garnet.

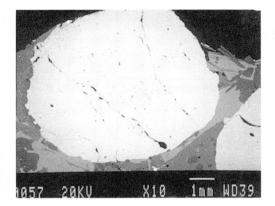

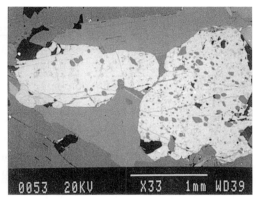

Figure 9.2 . fractured garnet crystal.

Figure 9.3. Garnet crystals with inclusions of quartz (grey) and other minerals (white).

The samples contained an average of 10% garnet, 7% kyanite, 33% quartz, 32% muscovite, minor amounts of phlogopite, albite and tremolite, and trace amounts of chlorite, amphibole, rutile, ilmenite, apatite, zircon, monazite, graphite and pyrrhotite. The garnet occurs as large, fractured, poorly formed crystals that are up to 13 mm in size (Figure 9.2). Most of the crystals are single crystals, although a few are twinned. The garnet contains numerous inclusions of quartz, and a few inclusions of rutile, apatite, chlorite, pyrrhotite, zircon, muscovite, tremolite, and plagioclase (Figure 9.3). Microprobe analyses show that the garnet is the almandine variety. The mineralogical study indicated that the garnet is of poor quality, but its suitability for abrasives would have to be tested.

9.6. QUARTZ

Quartz has many industrial uses, including metallurgical flux, foundry moulding, sandblasting, ceramics, and manufacturing such products as silicon, ferrosilicon, glass, glassfibre, silicon chips, optical fibres, etc. (Collings and andrews, 1989). The specifications of quartz for some applications are given in Table 9.8. Industrial quartz is usually obtained from vein quartz, massive quartz bodies, sand, sandstone and quartzite deposits. Mineral processing methods used to upgrade the quartz are screening, magnetic separations, gravitational separations, attrition scrubbing and washing, flotation, and acid leaching to remove the remaining iron minerals and carbonates. Some industrial operations, such as mining and sand blasting, produce large amounts of airborne quartz dust, which is a health hazard, particularly if large quantities of -10 μm quartz particles are inhaled (Burtan, 1984).

Mineralogical studies related to mineral processing of quartz and manufacturing of silicon products, are conducted to determine the properties of the quartz and the identities and textural relations of the impurities. In addition, airborne dusts in the workplace are commonly monitored by optical microscopy, SEM or XRD to determine the amount of quartz and its characteristics.

Table 9.8
Specifications* for some uses of quartz

Use	Size	SiO_2 (%)	FeO (%)	Al_2O_3 (%)	CaO (%)	MgO (%)	Ti ppm
Flux	0.5 - 2.5 cm	high	minor	minor			
Ferrosilicon	2 - 15 cm	>98	<1	<0.5	<0.2	<0.2	18
Silicon	2 - 15 cm	>99.5	<0.008	<0.4			18
Silica brick	- 2.5 cm	>95	<0.1	<0.1			
Glass, glass fibre	100 - 600 μm	>99	<0.025	<0.15	<0.15	<0.15	3
Silicon carbide	0.5 - 2 mm	>99	<0.1	<0.1			
Silicate chemicals	150 - 840 μm	>99	<0.1	<1.0	<0.25	<0.25	

* compiled from Collings and Andrews (1989) and Malvik and Lund (1990).

9.6.1. Mineralogical study of quartz

A mineralogical study of quartz from a high-grade deposit in Labrador is given here as an example of characterizing a quartz deposit to determine whether it is suitable for exploitation (Petruk et, al., 1992c). The average of chemical assay of 6 samples taken from the quartz body is, in wt %, SiO_2 99.4, FeO 0.17, Al_2O_3 0.15, CaO 0.15, MgO 0.08, K_2O 0.02, Na_2O 0.03, TiO_2 <0.02. The quartz is very pure and should meet specifications for many industrial uses without upgrading. Mineralogical studies show that the impurities in the quartz are hematite, magnetite, goethite, dolomite, magnesite, muscovite, biotite, chlorite, kaolinite, apatite, rutile, ilmenite, garnet, monazite, chalcopyrite and galena. Some of the iron oxides, mica and chlorite are relatively coarse-grained and the kaolinite is extremely fine-grained. These minerals occur along quartz grain boundaries (Figure 9.4), therefore, it should be possible to upgrade crushed and ground quartz products.

A grinding test was conducted to determine how the quartzite breaks with crushing and grinding. The combined sample was crushed to -1700 μm , then 2 kg samples of the crushed quartz were ground in a ball mill, using steel balls, for 5 minutes, 10 minutes and 15 minutes. The materials produced by each test were sieved. The results show that during crushing the quartz breaks largely into +600 μm particles, and during grinding it tends to break into two sizes ranges, -212 μm and +300 μm (Figure 9.5). The tests suggest that the quartz grains occur in two size ranges, -212 μm and + 300 μm.

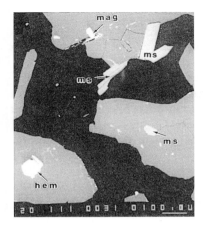

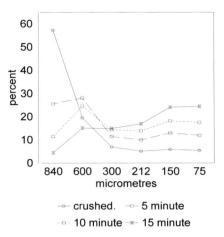

Figure 9.4. Impurities in quartz grains. Muscovite (ms), magnetite (mag), hematite (hem).

Figure 9.5. Size distributions of crushed and ground quartz.

9.6.2. Characterizing quartz with respect to manufacturing silicon and ferrosilicon

The manufacture of silicon and ferrosilicon involves feeding 2 to 15 cm pieces of quartz into a furnace. The operation requires a quartz that can withstand the thermal shock of being introduced into the furnace without disintegration or explosion. There is no satisfactory industrial laboratory method for determining the thermal properties of quartz, but Malvik and Lund (1990) described mineralogical properties that can indicate the suitability of the raw material. Some of the desirable properties of quartz used for manufacturing silicon and ferrosilicon are:

- A chemical composition that meets the specifications listed in Table 9.8.
- Fine-grained quartz. The linear expansion on heating quartz grains is less along the c-axis than along the a-axis, hence coarse-grained quartz has a lower thermal stability than fine-grained quartz.
- A quartzite mass that is composed of primary clastic grains in a silica cement matrix is the most desirable. Completely recrystallized quartzite, as evidenced by triple grain junctions, is undesirable. The cement matrix can be readily observed by cathodoluminescence.
- Mica is the most harmful trace mineral, but clay minerals, feldspar and pyrophyllite are also undesirable as they contribute the Al_2O_3 impurity. If the minerals occur in fissures and veinlets, they can be removed to a large extent by scrubbing and washing.

A dilatometry test is used to measure the expansion that takes place during heating of quartz. The test is performed on a small drill core, 3 cm in length and 0.9 cm in diameter. The core is heated and the expansion is measured. A normal expansion occurs at 573°C, which is the transformation temperature from α- to β-quartz, but there should be little to no expansion in the 750 to 965°C temperature range. Much of the expansion in this temperature range is caused by a mica impurity.

9.7. METHOD OF CHARACTERIZING AIRBORNE DUSTS

It is mandatory to monitor the airborne dust in many work areas to meet health and safety standards, because some mineral species, such as asbestos and quartz, are considered hazardous even when present in small quantities (Epstein, 1984; Burtan, 1984). Airborne dusts contain particles that are composed of a wide variety of minerals, which are dependent upon the site and source of the dust. The dusts are commonly collected on millipore filters (polycarbonate membrane filters) or glass slides, and are analyzed by either XRD (Knight et al., 1974) or with an optical microscope, scanning electron microscope (SEM) or/and transmission electron microscope (TEM). Some hazardous minerals, such as asbestos, may be present in such small quantities that it is difficult to detect them, much less identify them, by classical optical microscopy and XRD techniques (Skinner and Ross, 1994; Skinner et al., 1988; Merefield et al., 1995). The very small particles are usually analyzed manually with a SEM or a TEM. Some analytical techniques have been standardized for many work areas and are performed routinely. These techniques are either manual and time consuming or are designed to analyze for only one or two minerals.

An automatic technique for analyzing airborne dust samples was developed by Petruk and Skinner (1997) to detect and analyze particles that have been collected on polycarbonate membrane filters. The analysis is performed to:
- Identify the mineral in each particle by energy dispersive X-ray analysis (EDX).
- Measure the area (A), maximum diameter (D_{max}), aspect ratio (D_{max}/ave. particle width) and other morphological features of each particle. The average particle width is calculated as (A/D_{max}).
- Count the number of particles.

The technique uses an image analyzer interfaced with a SEM and EDX in such a manner that the image analyzer controls the operation of the SEM and can receive signals from the EDX. The analysis is performed by scanning each particle for a pre-selected period of time and sending the X-rays generated by each particle to EDX. The X-ray signal is sorted into X-ray counts for pre-selected elements which are known to be present in the minerals. The sorted data are then sent to the image analyzer for processing. When a particle has been scanned completely, the electron beam moves automatically to the next particle and scans it. When all the particles in a field of view have been analyzed completely the stage motors automatically move the sample to the next field of view. The image analysis system is programmed to analyze a pre-set number of fields of view in an unattended mode.

A dust sample mounted on a polycarbonate membrane filter was analyzed by scanning each particle for one second and using the elements Mg, Al, Si, K, Ca and Fe to identify the minerals. The mineral species were identified using a series of "if", "&", "*" and "/" statements that reflected ideal mineral compositions and were developed by a trial and error technique on known minerals at the selected analytical conditions. The values for the "if", "&", "*" and "/" statements are dependent upon the type and configuration of the EDX detector, the number of X-ray counts per second for each element, and the widths of each element window. The constraints defined by these statements need to be wide enough to allow for the compositional variations encountered by analyzing each particle a short period of time (i.e. 1 second), and narrow enough to reject other minerals from the designated mineral bin.

The sample was scanned for two hours at a magnification of 300 times. Many fields of view did not contain any particles, but some fields contained up to 15 particles. The minerals identified in the dust sample were lizardite, anthophyllite, tremolite, talc, biotite, MgFe silicate, MgCa silicate, MgFeCa silicate, Al silicate, Ca silicate, Café silicate, Carbonates (calcite and dolomite) and quartz. A total of 1966 particles were analyzed. The aspect ratios varied from less than 3 to greater than 5 for all minerals, and to greater than 10 for lizardite (asbestos), tremolite and talc. In particular about 71 to 83% of all minerals except the MgFe silicate had aspect ratios less than 3. Between 1 and 2 % of the lizardite (asbestos), tremolite and talc had aspect ratios greater than 10. All the MgFe silicate particles had aspect ratios of less than 3. The particle lengths of the minerals ranged from 0.8 to 43 μm; but, the average particle lengths of the different minerals varied between 1.8 and 4.9 μm.

CHAPTER 10

APPLIED MINERALOGY TO TAILINGS AND WASTE ROCK PILES - SULFIDE OXIDATION REACTIONS AND REMEDIATION OF ACIDIC WATER DRAINAGE

10.1. INTRODUCTION

Metal mining operations produce large volumes of tailings and waste rock, which need to be disposed of at nearby locations. The wastes usually contain small to large amounts (<1 to ~60 %) of pyrite and/or pyrrhotite and trace amounts of valuable minerals. Oxidation of pyrite and pyrrhotite generates acidic water, which dissolves the sulfide minerals and releases hazardous elements (e.g. Cu, Zn, Pb, Ni, Cd, As, Sb, Se, Cr, Co, Hg, Mo, U, etc) into the water drainage system. The acidic pore water in the tailings pile is initially neutralized by lime and calcite, and to some degree, by the other minerals that are present in the tailings and waste rock piles. The pore water, however, becomes acidic when the lime and calcite are depleted (Jambor and Blowes, 1998). Under acidic conditions, decomposition of minerals proceeds and oxidation of sulfides is enhanced by bacterial action (Gould et al., 1994).

Mine development has always required a plan for disposal of tailings and waste rock. In the past, however, there was little concern for site rehabilitation and potential release of hazardous effluents. Consequently tailings and waste rock piles were abandoned at the end of a mining operation. The outflow of hazardous effluents into the environment led to major changes, and since 1970 abandonment without rehabilitation has been prohibited in North America (Jambor and Blowes, 1998). Current planning for disposal of tailings and waste rock involves major environmental assessments.

Because acid mine drainage (AMD) and acid rock drainage (ARD) produce environmental hazards, numerous investigations have been conducted on AMD and ARD at CANMET, Department of Energy, Mines and Resources, Canada; at the Waterloo Centre for Groundwater Research, University of Waterloo, Waterloo, Ontario, Canada; at other institutes and by industry (Robertson, 1994; Jambor, 1994; Alpers et al., 1994; Blowes and Ptacek, 1994; Feasby and Tremblay, 1995; Gould et al., 1994; vanHuyssteen, 1998a; Paktunc et al., 1998; Paktunc and Davé, 1999; Paktunc, 1999a, 1999b; Mend, 1989, 1991, 1992). Parallel investigations have been conducted in other countries (Ritchie, 1994a, 1994b; Bigham, 1994). The investigations were carried out by studying abandoned tailings piles and/or by conducting laboratory tests using lysimeters and/or columns. Lysimeters and columns are containers charged with tailings, and the charge is exposed to simulated weathering conditions. The reactions within the lysimeters and/or columns are measured periodically. In terms of scientific progress, studies on AMD and ARD are young, but much information has been compiled and interpreted nonetheless.

A proper understanding of the mineralogical and chemical reactions within tailings and waste rock piles is essential when planning for disposal of mine wastes in an environmentally acceptable manner. Such an understanding requires integration of knowledge on pore water, oxygen pressures, hydraulic pressures, chemistry, mineralogy and other factors. Thus, collaboration between hydrogeochemists and mineralogists is essential. For example, the policy of British Columbia in Canada stipulates that *"Where there is a potential for the generation of acid drainage or metal release through weathering or dissolution, the proponent should determine the range, variability, and central tendencies for the following properties: elemental composition; mineralogy; readily soluble constituents; sulfide types (amount, reactivity, and spatial distribution); carbonate types (amount, reactivity, and spatial distribution); and mineralogical rock-fabric characteristics that will influence weathering."* (Price and Errington, 1994; Jambor and Blowes, 1998).

The mining industry and environmental regulatory bodies rely on various tests that can be performed routinely, relatively inexpensively and rapidly by chemical analysis (Jambor and Blowes, 1998). A procedure utilized in the pre-mining environmental assessment, and in waste management plans for closure, is the determination of an acid-base accounting. The objective is to analytically estimate the quantities of minerals capable of generating acid and the quantities of minerals that may neutralize the acid during the weathering process (Mend, 1991). The difference between the acid producing potential (AP) and acid neutralizing potential (NP) is the net neutralization potential (NNP). The AP value is usually taken as the sulfur content determined by chemical analysis of the sample. The NP value is commonly obtained by boiling a sample in a known quantity of HCl, and determining how much of the acid has been consumed. The AP and NP data may be subject to error because the values do not measure the mineralogy of the rock even though minerals are the sources for the AP and NP data. For example, the rock or tailings pile may contain sulfates, such as barite and gypsum, that do not contribute to acid generation. Similarly the test for determining the NP value does not consider neutralizing minerals that are not soluble in HCl.

As remediation and control of acid mine drainage from tailings and waste rock piles are dependent upon reactions in the piles, and as the minerals in the piles are the source materials for the reactions, this chapter discusses reactions in tailings and waste rock piles, with emphasis on the role of mineralogy in understanding the reactions. Although most of the discussions are on reactions in tailings, which consist of unwanted material discarded during mineral processing operations, similar reactions occur in waste rock piles, which are wall rock material removed to access and mine the ore.

10.2. GENERAL CHARACTERISTICS OF TAILINGS AND WASTE ROCK PILES

Blowes and Ptacek (1994) and others have shown that typical tailings and waste rock piles (Figure 10.1) consist of:
(1) a vadose zone at the top section of the pile, which is the zone of sulfide oxidation and acid generation,
(2) a capillary zone below the vadose zone and just above the water table, which is a zone of acid neutralization and chemical precipitation,
(3) a saturated zone below the water table, which is the zone of attenuation and dissolution,
(4) a zone of transport of dissolved species at the toe of the pile and in ground water beneath the pile.

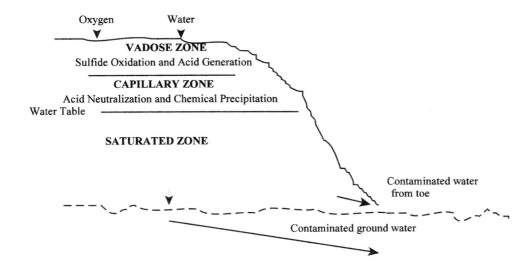

Figure 10.1. Schematic diagram of the zones in tailings and waste rock piles.

The oxidation of sulfides and generation of acidic water occurs largely in the vadose zone (Ritchie, 1994a, 1994b). The rate of oxidation is controlled by the amount of moisture and oxygen in the pore spaces in the tailings and waste rock piles. A relatively small quantity of water is required for the reaction to proceed, hence pyrite oxidation occurs even in arid regions (Ritchie, 1994a). During dry seasons the pyrite may continue to oxidize even when there is not enough water from rainfall to wash away the dissolved ions. The released ions are subsequently flushed out during rainy seasons. In contrast, the rate of pyrite oxidation decreases as the availability of oxygen decreases. The oxygen pressure within tailings piles is relatively low, and decreases with depth. Consequently the oxidation reaction is slow, and it would take several centuries to oxidize all the pyrite in a tailings pile about 10 m high.

Studies on remediation of acid generation from tailings piles have shown that the rate of pyrite oxidation and production of acidic water can be reduced by decreasing the flow of oxygen into the tailings pile. This can be partly achieved with:
- an impervious cover over the tailings pile,
- a high water table in the tailings pile,
- underwater disposal of tailings (Robertson, 1994).

A high water table reduces sulfide oxidation because the rate of diffusion of atmospheric oxygen is decreased by more than three orders of magnitude in the saturated zone below the water table. Constructing impermeable dams around the tailings impoundment can raise the level of the water table within the tailings impoundment. It has also been found that discharging the tailings at the centre of an impoundment can temporarily increase the height of the water table (Robertson, 1994).

Underwater disposal of tailings minimizes, and may prevent, sulfide oxidation and its attendant acid generation (Feasby et al., 1997; Roberstson et al., 1997). For example, a column test on pyritic uranium tailings submerged under shallow water showed negligible acidity, and only a trace of oxidation activity was detected after seven years (Paktunc and Davé, 1999b). vanHuyssteen (1998a) studied a pyrrhotite-rich tailings from a nickel ore that was deposited in a lake, but the top of the tailings pile was about 1.5 m above the lake surface. Some of the pyrrhotite oxidation, which occurred above the lake surface was sustained below the water table. In particular, native sulfur and the secondary minerals, szomolnokite and natro-jarosite, were found 15 m below the surface of the tailings pile. These secondary minerals were obviously transported below the water table, hence a tailings pile must be totally submerged to minimize or prevent oxidation.

Disposal of mine wastes may include underwater disposal, or involve site preparation by constructing impermeable impoundments or retaining dams across topographic depressions. In environmentally sensitive areas the entire impoundment may be lined with a low permeability clay or synthetic membrane to minimize seepage. Furthermore, facilities for collecting samples that are used to monitor reactions are commonly installed while constructing the waste impoundment (Jambor and Blowes, 1998).

10.3. MINERALOGICAL STUDIES OF TAILINGS AND WASTE ROCK PILES

10.3.1. Objective

The primary objective of a mineralogical study of tailings and waste rock piles is to provide input parameters for a hydrogeochemical and acid water remediation model by characterizing the starting materials and the oxidation solids.

The requirements from a mineralogical study were outlined by Jambor and Blowes (1998) and Price et al. (1997). They include:
- identification of potential acid generation and metal sources, with emphasis on sulfide mineralogy,
- determination of potential contribution of sulfur to the S assay by sulfur-bearing minerals, such as barite, which do not generate acid,
- identification of potential neutralization sources, with emphasis on carbonate mineralogy and slow release alkaline aluminosilicates,
- evaluation of the most reactive acid-generation and neutralization sources (sulfides and carbonates) and their potential to occur preferentially along fracture planes and in the finer-size fraction, where they are available to contribute to geochemical reactions,
- identification of readily soluble constituents,
- identification of mineralogical or rock-fabric characteristics that will influence weathering.

10.3.2. Sampling and method of analysis

The following is a synthesis of procedures that were described by Jambor (1994) and Jambor and Blowes (1998) for collecting and analysing tailings and waste rock samples.
The mineralogist would obtain a subset of representative samples that were taken as cores through the mine wastes pile. For investigations of a tailings impoundment, the cores should

be taken by a hydrogeochemist. The sampling sites should be designed to give a good spatial distribution, as well as proximity to piezometer sites, so that mineralogical and hydrogeochemical results could be correlated and integrated. Waste rock piles should be sampled by the mineralogist because he is best trained to notice subtle oxidation effects and development of secondary minerals, all of which should be sampled because they indicate the nature of the ions in solution and the solid-phase controls on their distribution.

A simple method of taking cores of tailings piles is to drive a small diameter (5 - 7.5 cm) PVC or thin-walled aluminum pipe through the tailings. Core retention can be accomplished by inserting an iris-diaphragm-type cup at the entry point. To minimize oxidation and other chemical reactions, and to prevent loss of pore waters, cores should be end-sealed and frozen at the minesite. Freezing is essential if the cores are to be extricated intact, preferably by cutting the barrel, and its contained core, along length with a band saw.

The cores would be logged in the laboratory and samples should be judiciously selected for study from the tens or hundreds of metres of core that may be available. Logging will be facilitated by extruding the frozen cores unto plastic trays. Use of plastic trays is important because the cores are commonly saturated with pore waters from which tertiary precipitates will crystallize as drying proceeds. The cores should be logged upon thawing, during drying, and after drying, which takes about 1 week. The logging is performed to note the colour, mineral variations, sedimentary structures, and sedimentary cementation, so that distributions of tertiary precipitates can be recorded. Forced drying such as low temperature drying is not recommended because the mineralogy of the precipitates can be severely affected.

The vadose zone encompasses the oxidized layers at the top of the impoundment and may extend to the hardpan layer and part of the reduction zone (Boorman and Watson,1976; Blowes et al., 1992). This zone will thus have the largest variations in mineralogy, and sampling of it should be more detailed than of the underlying saturated zone below the water table. A series of samples is removed intact for microscopical examination, and companion samples are taken for related studies such as X-ray diffractometry and chemical analyses. Using intact samples minimizes losses of very fine-grained secondary minerals, and retains textural relationships of the primary and secondary minerals. Dried tailings samples are commonly poorly cemented and difficult to handle without disintegration. To facilitate mineralogical studies the samples are normally impregnated with a resin; preferably a low-viscosity resin such as a cyanoacrylic adhesive (Lastra and Greer, 1992). To prevent dissolution of secondary and tertiary minerals, polished sections and polished-thin sections are prepared either dry, or without aqueous lubricants.

The standard procedure for conducting mineralogical studies of rocks and tailings is to obtain a polished-thin section and an X-ray diffractogram of each selected sample. Transmitted light microscopy is used to obtain a petrographic description, and reflected light microscopy to define the "ore" mineralogy. The microscopical study need not be exhaustive, but there must be a correlation with the X-ray diffractometry results for the minerals identified. On the other hand, it must be recognized that tailings deposits may not be homogeneous. The microscopy will reveal the presence of minerals too sparse or poorly crystalline to be determinable in the X-ray diffractogram. Scanning electron microscopy

(SEM) and microprobe (MP) analyses are commonly used to supplement the information obtained by optical microscopy and X-ray diffractometry. Some grains cannot be identified by the composition determined with the MP. For example, a quantitative MP analysis of an Fe-oxide alteration rim will not unequivocally identify the rim as goethite, lepidocrocite, ferrihydrite, schwertmannite, or mixtures of some of these and possibly other minerals. The MP identification would have to be corroborated by other methods. If the grains are large enough, they can be extracted from the polished-thin section and collected on an ultra-slim glass fibre that is coated with an adhesive. The mineral powder on the glass fibre would be identified by X-ray diffraction.

Image analysis and quantitative x-ray diffractometry may be used to determine mineral quantities.

10.4. MINERALS

Jambor (1994) classified the minerals in tailings samples as follows:
- primary minerals: minerals that were deposited in the impoundment,
- secondary minerals: minerals that formed within the impoundment,
- tertiary minerals: minerals that crystalize from pore waters after the samples have been removed from their source (e.g., during core drying),
- quaternary minerals: late-stage oxidation products that form during storage of dried samples.

Table 10.1 lists the minerals that have been observed as secondary, tertiary and quaternary minerals in tailings piles. The minerals gunningite, bianchite and boyleite, listed in the table, were observed only as tertiary minerals.

10.5. REACTIONS IN TAILINGS PILES

The rate of sulfide oxidation and release of dissolved metals is greatest shortly after tailings deposition ends because oxygen can diffuse rapidly across the short distance between the tailings surface and the depth of active sulfide oxidation (Alpers et al., 1994; Blowes and Ptacek, 1994; Ritchie, 1994a, 1994b). As the near-surface sulfide minerals are depleted, the zone of sulfide oxidation moves deeper into the pile, and the rate of oxidation slows as the length of the oxygen diffusion path increases. Oxygen diffusivity is reduced further as barriers such as hard-pan are produced, and decreases even more as the moisture content in the tailings pile increases. In the water saturated zone below the water table the rate of sulfide oxidation is several orders of magnitude slower than in the vadose zone (Blowes and Ptacek, 1994).

10.5.1. Oxidation of sulfide minerals

Jambor (1994) suggested that the relative resistance of sulfide minerals to oxidation in tailings environments increases in the order: *pyrrhotite - (sphalerite-galena) - (Pyrite-arsenopyrite) - chalcopyrite.* The high resistance of chalcopyrite to oxidation was observed in the tailings impoundment at Waite Amulet, Quebec by the presence of chalcopyrite in the near-surface oxidized zone where all other sulfides had been consumed (Blowes and Jambor, 1990; Petruk and Pinard, 1986). Similarly slow rates of dissolution were observed for chalcopyrite and for

Table 10.1.
Secondary, tertiary and quaternary minerals observed in sulfide-rich tailings piles*

Iron oxides, Oxyhydroxides		Sulfates (cont'd)	
goethite	α-FeO(OH)	barite	$BaSO_4$
lepidocrocite	γ-FeO(OH)	epsomite [ter.]	$MgSO_4.7H_2O$
akaganéite	β-FeO(OH,Cl)	hexahydrite[ter.]	$MgSO_4.6H_2O$
maghemite	γ-Fe$_2$O$_3$	pentahydrite[ter.]	$MgSO_4.5H_2O$
hematite	Fe_2O_3	starkeyite [ter.]	$MgSO_4.4H_2O$
ferrihydrite	$Fe_2O_3.9H_2O$	anglesite	$PbSO_4$
Sulfates		thenardite [ter.]	Na_2SO_4
gypsum	$CaSO_4.2H_2O$	alunogen [ter.]	$Al_2(SO_4)_3.17H_2O$
bassanite	$2CaSO_4.H_2O$	unidentified	$(Cu,Fe)(SO_4,AsO_4,PO_4)$
jarosite	$KFe_3(SO_4)_2(OH)_6$	copiapite	$Fe^{2+}Fe^{3+}_4(SO_4)_6.20H_2O$
hydronium jarosite	$(H_3O)Fe_3(SO_4)_2(OH)_6$	antlerite	$Cu_3(SO_4)(OH)_4$
natrojarosite	$NaFe_3(SO_4)_2(OH)_6$	brochantite	$Cu_4(SO_4)(OH)_6$
schwertmannite	$Fe_8O_8SO_4(OH)_6$	alunite	$KAl_3(SO_4)_2(OH)_6$
fibroferrite	$Fe(SO_4(OH).5H_2O$	jurbanite	$Al(SO_4)(OH).5H_2O$
melanterite [ter.]	$Fe^{2+}SO_4.7H_2O$	basaluminite	$Al_4(SO_4)(OH)_{10}.5H_2O$
ferrohexahydrite [ter.]	$Fe^{2+}SO_4.6H_2O$	Other minerals	
siderotil [ter. Quat.]	$Fe^{2+}SO_4.5H_2O$	marcasite	FeS_2
rozenite [ter. Quat]	$Fe^{2+}SO_4.4H_2O$	covellite [ter.]	CuS
szomolnokite	$Fe^{2+}SO_4.H_2O$	sulfur	S
goslarite [ter.]	$ZnSO_4.7H_2O$	cristobalite	SiO_2
gunningite [ter.]	$ZnSO_4.H_2O$	vermiculite	$(Mg,Fe,Al)_3(Si,Al)_4(OH)_2.4H_2O$
bianchite [ter.]	$ZnSO_4.6H_2O$	smectite	$X_{0.3}Y_{2-3}(Si,Al)_4O_{10}(OH)_2.nH_2O$
boyleite [ter.]	$ZnSO_4.4H_2O$	kaolinite	$Al_2Si_2O_5(OH)_4$
ashoverite	$Zn(OH)_2$	erythrite	$Co_3(AsO_4)_2.8H_2O$

* from (Jambor and Blowes, 1998; Alpers et al., 1994; Blowes et al., 1991; vanHuyssteen, 1998a).
[ter.] = observed as secondary and ternary minerals (Jambor, 1994).
[Quat.] = observed as quaternary minerals (Jambor, 1994).

sphalerite in the Heath Steele impoundment in New Brunswick where the tailings pore waters locally had a pH = <3 (Blowes et al., 1991, 1992; Jambor, 1994). The slow dissolution rate for sphalerite shows that the dissolution rate for some sphalerite varieties is as slow as for chalcopyrite.

10.5.1.1. Pyrite

Blowes and Ptacek (1994) reported that the reaction of pyrite in tailings and waste rock piles is:

$$FeS_2 + 7/2\ O_2 + H2O \Rightarrow Fe^{2+} + 2SO_4^{2-} + 2H^+ \tag{1}$$

Subsequent oxidation of Fe^{2+}, and hydrolysis and precipitation of $Fe(OH)_3$, produce an additional two moles of H^+, resulting in the overall reaction:

$$FeS_2 + 15/4\ O_2 + 7/2H2O \Rightarrow Fe(OH)_3 + 2SO_4^{2-} + 4H^+ \tag{2}$$

Reaction (1) shows that oxidation of pyrite produces Fe^{2+}, SO_4^{2-} and H^+. The dissolved Fe^{2+} may (a) remain in solution, (b) precipitate as ferrous sulfate and ferrous hydroxide minerals, or (c) oxidize and precipitate as ferric oxyhydroxides. The dissolved Fe^{2+} may be carried along in the water, giving it the appearance of a sludge. The dirty sludge-like water is ultimately discharged into the surface-water flow system, where Fe^{2+} is oxidized to Fe^{3+} and precipitates as ferric oxyhydroxides (e.g. goethite, lepidocrocite, etc.) (Blowes and Ptacek, 1994). The secondary minerals (e.g. ferrous sulfates, ferrous hydroxide and ferric oxyhydroxides) commonly precipitate as coatings on pyrite, pyrrhotite and other minerals (Figure 10.2, 10.3, 10.4). Some of the secondary minerals are relatively soluble and may re-dissolve in the tailings pore water.

 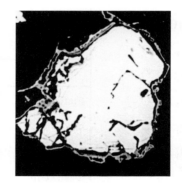 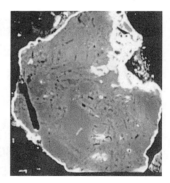

Figure 10.2. Fe sulfate border on pyrrhotite. Waite Amulet tailings pile.

Figure 10.3. Goethite border (light grey) on magnetite. Waite Amulet tailings pile.

Figure 10.4. Goethite border (White) on silicate particle. Waite Amulet tailings pile.

A model for initiation of acid mine drainage has been suggested from work at CANMET and University of Waterloo (Gould, 1997). At high pH values pyrite slowly oxidizes to goethite and elemental sulfur (Nesbitt and Muir, 1994). Neutrophillic thiobacilli bacteria, which grow at a pH of 6 to 8.5, oxidize the elemental sulfur to produce sulfuric acid. The sulfuric acid reduces the pH and clears the mineral surface so that chemical oxidation of the mineral can continue. When the pH has been lowered to the 4 to 4.5 range acidophilic bacteria, such as thiobacillus ferrooxidans, oxidize the pyrite, the aqueous Fe^{2+}, the elemental sulfur and the sulfates and generate acid. (Gould et al., 1994; Bigham, 1994). At a pH of <3 the activity of Fe^{3+} becomes significant because the Fe^{3+} replaces O_2 as the primary oxidant. At pH <2.5, a near-steady-state cycling of Fe occurs via the oxidation of primary sulfides by Fe^{3+} and the subsequent bacterial oxidation of regenerated Fe^{2+} (Kleinmann et al., 1981; Gould et al., 1994; Bigham, 1994).

The SO_4^{2-} generates acidic pore waters, which dissolve other minerals and release deleterious elements into the pore waters. The loaded acidic water escapes from the toe of the tailings pile into the surface drainage system, and into the soil and groundwater beneath the tailings and rock piles.

The H^+ produces acid-neutralizing reactions when it comes in contact carbonates, hydroxides and other base containing solids (Blowes and Ptacek, 1994). The acid-neutralizing reactions increase in the pore-water pH, which is often accompanied by precipitation of metal-bearing hydroxide and hydroxysulfate minerals, commonly as coatings on other minerals. The precipitation of inhibitory mineral coatings on sulfides, especially pyrite, can decrease the rate of sulfide oxidation.

10.5.1.2. Pyrrhotite
The oxidation of pyrrhotite is faster than oxidation of pyrite. The reaction of pyrrhotite is:

$$Fe_{(1-x)}S + (2-x/2)O_2 + xH_2O \Rightarrow (1-x)Fe^{2+} + SO_4^{2-} + 2xH^+ \quad (3)$$

or may proceed to partial completion, generating Fe^{2+} and elemental S^0 through the reaction:

$$Fe_{(1-x)}S + (2-2x)Fe^{3+} \Rightarrow (3-3x)Fe^{2+} + S_{(S)}^{0} \quad (4)$$

In the initial stages of alteration, pyrrhotite is replaced by marcasite that is heterogeneous and somewhat fibrous in appearance, which is a texture related to the basal parting of the host pyrrhotite (Figure 10.5). With more advanced alteration the marcasite is replaced by Fe oxyhydroxides, commonly as rims of goethite, less commonly as lepidocrocite, and subsequently by pseudomorphs of goethite (Jambor and Blowes, 1998). Elemental sulfur(S^0) may be present as an intermediate product in the oxidation of pyrrhotite, and has been found in several tailings piles (Blowes and Ptacek, 1994).

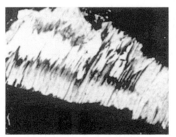

Figure 10.5. Pyrrhotite (white at centre of particle) largely replaced by a fibrous-like marcasite.

10.5.2. Dissolution of Carbonate minerals

The stability, or resistance to alteration, of carbonate minerals is important because of the role of these minerals in neutralizing acidic drainage. Reactions involving carbonate mineral dissolution in mine wastes have the potential of maintaining the near neutral pH conditions that are required to prevent the metal pollution that occurs by dissolving other sulfide minerals (Ritchie, 1994). The carbonate minerals in the rocks associated with ore deposits are almost invariably calcite ($CaCO_3$), dolomite ($CaMg(CO_3)_2$), siderite ($FeCO_3$) and ankerite ($Ca(Fe,Mg)(CO_3)_2$. Field observations show that the order of increasing resistance to dissolution is *calcite - dolomite - ankerite - siderite* (Jambor and Blowes, 1998). The rate of calcite dissolution is sufficient to maintain near-equilibrium conditions, and to maintain pore water pH in the range of 6.5 - 7.5. Laboratory column experiments, conducted for 7.5 years, showed that pyritic uranium tailings amended with sufficient fine-grained limestone did not produce acid mine drainage (Paktunc and Davé, 1999). In contrast, the same tailings amended with the same weight proportion of coarse-grained limestone produced some acidic drainage, and without limestone the tailings oxidized rapidly producing highly acidic drainage.

Paktunc (1999b) proposed a model for determining the acid neutralizing potential (NP) values by using mineralogical data. He determined that, as a rough guide, waste material must contain at least 12 times as much calcite as pyrite (in volume %), or 8 times as much calcite as pyrhottite (in volume %), before it can be considered as having no acid mine drainage potential.

Calcite-dissolution reaction follows the general form:

$$CaCO_3 + H^+ \Rightarrow Ca^{2+} + HCO^{3-} \tag{5}$$

This reaction consumes calcite and H^+, releases dissolved cations (e.g. Ca and HCO^{3-}) to the tailings pore water, and increases the pore-water alkalinity concentrations. The HCO^{3-} may combine with Fe^{2+} and lead to saturation or supersaturation of Fe carbonates and precipitation of secondary siderite (Morin and Cherry, 1986; Blowes et al., 1991; Alpers et al.,1994). The dissolved Ca and SO_4 may combine to form gypsum, which precipitates as part of hardpan layers (Blowes et al., 1991).

Following depletion of calcite, the pore-water pH drops abruptly to near 4.8 (Figure 10.6), and siderite begins to dissolve. At this moderate pH condition precipitation of many dissolved metals is favoured resulting in continued removal of elements from the water. In the calcite- and siderite-buffered zones precipitation of metal hydroxides or hydroxysulfate, including gibbsite, amorphous $Al(OH)_3$, amorphous $Fe(OH)_3$, ferrihydrite, goethite, and schwertmannite, is favoured, leading to accumulations of these minerals as cements or grain coatings. The Fe is derived from the sulfide oxidation reactions, and the dissolved Al is derived from dissolution of aluminosilicates (Blowes and Ptacek, 1994).

10.5.3. Dissolution of aluminum hydroxides, aluminosilicates and silicates

As acid generation continues and the carbonate minerals are depleted, the pH drops abruptly until the dissolution of the next pH buffer, $Al(OH)_3$, is favoured (Figure 10.6). Dissolution of

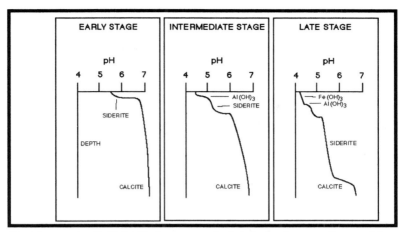

Figure 10.6. Development of pH buffering zones during early, intermediate and late stages of sulfide oxidation (from vanHuyssteen, 1998b).

$Al(OH)_3$ buffers the pH to values between 4.0 and 4.3. When $Al(OH)_3$ is consumed the pH drops again, favouring dissolution of $FeOH_3$, and resulting in pH values that fall below 3.5. When the dolomite, siderite and Al oxides are used up the pH of the pore water drops to around 2 - 4 and the primary sulfides, such as chalcopyrite, galena, sphalerite, arsenopyrite, etc., dissolve and release deleterious elements including Cu, Pb, Zn, As, Se, Sb. etc. (Blowes and Ptacek, 1994).

Under very low pH conditions, after all carbonates and simple hydroxides are depleted, the dissolution of aluminosilicates becomes an important acid-neutralizing mechanism. Evidence of aluminosilicate dissolution is an increase in the Si and Al contents in the pore water. Dissolution of aluminosilicate minerals may result in depletion of the original mineral and formation of a second more stable mineral (Blowes and Ptacek, 1994).

Silicate minerals are much less amenable to dissolution than carbonates, but the dissolution rates in tailings piles may be significant. The dissolution rate for silicate minerals decreases in the order *forsterite - pyroxenes - biotite - albite - muscovite - quartz* (Jambor and Blowes, 1998). The dissolutions of these minerals is responsible for the concentrations of major ions such as Mg, Na, K, Al, and sulfate in the drainage water. In general, the pH of the drainage water decreases with time as various reactions proceed within the waste-rock dump until pH reaches 2.0 - 4.0 (Ritchie, 1994a, 1994b).

Methods of calculating dissolution rates for minerals are discussed by Paktunc (1999a). He reported the order of decreasing dissolution rates for rock forming minerals at pH 5 as: *brucite - calcite - olivine (Fo_0) - dolomite - spodumene - plagioclase (An_{100}) - wollastonite - nepheline - olivine (Fo_{91}) - jadeite - olivine (Fo_{100}) - glaucophane - diopside - enstatite - hornblende - augite - plagioclase (An_{80}) - serpentine (antigorite) - plagioclase (An_{40}) - serpentine (chrysotile) - gibbsite - biotite - microcline - plagioclase (An_0) - plagioclase (An_{20}) - sanidine - talc - K-feldspar - phlogopite - muscovite - chlorite - epidote - kaolinite - quartz - montmormillonite - anthophyllite.* Paktunc (1999a) also reported the order of decreasing dissolution rates for rock forming minerals at pH 2 as: *calcite - dolomite - plagioclase (An_{100}) - spodumene - fayalite - glaucophane*

- wollastonite - olivine (Fo$_{91}$) - jadeite - olivine (Fo$_{100}$) - magnetite - diopside - hornblende - biotite - plagioclase (An$_{80}$) - augite - enstatite - plagioclase (An$_{40}$) - plagioclase (An$_{20}$) - plagioclase (An$_{60}$) - K-feldspar - muscovite - plagioclase (An$_{20}$) - kaolinite - anthophyllite.

10.5.4. Precipitation of secondary minerals

Secondary minerals precipitate throughout the vadose zone in the tailings pile, some in interstitial spaces and some as an armour on mineral particles. The secondary minerals have various degrees of solubility; some precipitate at one set of conditions in the tailings pile and redissolve at another set of conditions. Other secondary minerals are practically insoluble and do not redissolve. In arid or semi-arid environments, metal-bearing soluble sulfate minerals precipitate during dry periods and dissolve during wet periods, releasing the metals and causing dramatic seasonal variations in metal concentrations (Alpers et al., 1994).

Precipitation and dissolution of the sulfate-bearing secondary minerals melanterite, gypsum, and anglesite are rapid and seem to limit the concentrations of Fe^{2+}, Ca, Pb and SO$_4$ in the pore waters . Melanterite is a soluble sulfate mineral and high concentrations of Fe^{2+} are observed in pore waters where melanterite is present (Alpers et al., 1994). Gypsum is less soluble, and concentrations of dissolved Ca in pore waters are consistently <800 mg/l. Anglesite is relatively insoluble and relatively low concentrations of dissolved Pb (<20 mg/l) are maintained by the anglesite solubility. Below the hardpan layer the carbonate content of the tailings pore-water approaches saturation with respect to siderite (Ptacek and Blowes, 1994), and the mineral begins to precipitate. In such cases both primary and secondary siderite may be present.

Precipitation and dissolution reactions can have both beneficial and detrimental effects in mine wastes. Precipitation of relatively insoluble minerals, such as anglesite, can maintain low concentrations of dissolved metals in the tailings pore water while increasing the mass of the metal accumulates in the tailings solids. After the most intense period of sulfide oxidation and dissolved metal release has passed, and the pH has dropped to 2- 4, the secondary precipitates will dissolve and contribute dissolved metals to the tailings pore-waters for long periods of time (Alpers et al., 1994).

As the solubility of the secondary minerals plays an important role in the behaviour of metals in the tailings piles, the characteristics of the secondary minerals are discussed below with respect to solubility.

10.5.4.1. Soluble iron sulfates

Melanterite (Fe^{2+}SO$_4$.7H$_2$O) is probably the most common soluble Fe sulfate. It forms by combining Fe^{2+} and SO$_4{}^{2-}$ by a process that includes evaporation. It is pale glue green and occurs as stalactites in open mine tunnels (Alpers and Nordstrom, 1991) and as pore-filling cement in the hardpan layers of sulfide tailings (Blowes and Jambor, 1990). Its stability range is defined by temperature and water activity (relative humidity). With increasing temperature and /or decreasing water activity melanterite may dehydrate to **rozenite (Fe^{2+}SO$_4$.4H$_2$O)** (Alpers et al., 1994). The presence of solid solution substitutions can also affect the product. Jambor and Traill (1963) noted that, under identical conditions, copper-free melanterite dehydrated to rozenite, but copper-bearing melanterite dehydrated to **siderotil (Fe^{2+}SO$_4$.5H$_2$O)**. The divalent ions such as Cu, Zn, Ni, Mg and Mn have been incorporated as solid solutions in melanterite. In particular, a stalactite from Richmond mine at Iron Mountain has the composition: $(Fe_{0.534}Zn_{0.281}Cu_{0.142}Mg_{0.043})SO_4.7H_2O$ (Alpers et al., 1994).

10.5.4.2. Less soluble sulfate minerals

The less soluble sulfate minerals are alunite $(KAl_3(SO_4)_2(OH)_6)$, jarosite $(KFe_3(OH)_6(SO_4)_2)$, schwertmannite $(Fe_8O_8(OH)_6SO_4)$, jurbanite $(Al(SO_4)(OH).5H_2O)$, basaluminite $(Al_4(SO_4)(OH)_{10}.5H_2O)$, antlerite $(Cu_3(SO_4)(OH)_4)$, brochantite $(Cu_4(SO_4)(OH)_6)$ (Alpers et al., 1994), barite $(BaSO_4)$, anglesite $(PbSO_4)$ (Blowes et al., 1991), an unidentified amorphous poorly crystalline Al hydroxy-sulfate (Nordstrom et al., 1984), and solid solutions of $RaSO_4$ in alkali earth sulfate minerals. The solid solutions of $RaSO_4$ in alkali earth sulfate minerals, including barite $(BaSO_4)$ and celestite $(SrSO_4)$, provide an important mechanism for the attenuation of radium, which carries a significant amount of radiation in wastes from uranium mining (Langmuir and Melchior, 1985).

10.5.4.3. Insoluble ferric oxyhydroxides and sulfates

Goethite (α-FeO(OH) is a stable form of Fe^{3+} oxide, and contains minor to trace amounts of other elements. It is commonly a minor to trace constituent, and occasionally a major constituent, in mine-drainage precipitates and occurs as small yellowish-brown short particles (Brady et al., 1986). XRD line-widths at half height commonly correspond to a particle size of about 15 nm. The Mössbauer spectrum consists of a doublet, indicating superparamagnetic relaxation that is probably a result of small particle size (Murad et al., 1994).

Lepidocrocite (γ-FeO(OH) is a polymorph of goethite that is commonly recognizable by bright orange colours (Schwertmann, 1993). The mineral is uncommon in mine tailings, but Blowes et al. (1991) and Jambor (1994) observed lepidocrocite as a cementing agent in hardpans, and Milnes et al. (1992) detected poorly crystalline lepidocrocite in an ocherous sludge collected from a retention pond associated with a uranium mine in Australia.

Ferrihydrite ($Fe_2O_3.9H_2O$) is an often misused synonym for amorphous ferric hydroxide. The mineral is rusty-reddish brown, and usually occurs as aggregated spherical particles 2 to 6 nm in diameter (Schwertmann, 1993; Bigham, 1994). Ferrihydrite is usually poorly crystallized but the XRD patterns display 2 to 4 broad peaks (Carlson and Schwertmann, 1981).

Schwertmannite ($Fe_8O_8SO_4$ (OH)$_6$) is the most common mineral associated with ocherous precipitates from acidic sulfate waters. It is bright yellow, fine-grained and poorly crystallised, displaying eight broad peaks in the XRD pattern (Bigham, 1994).

Jarosite ($KFe_3(SO_4)_2(OH)_6$) is a common mineral in vadose zones of mine tailings (Fanning et al., 1993), usually appearing as straw coloured mottles and efflorescences. It is usually well crystalized and easily identified from its characteristic XRD pattern. Jarosite is a member of an iron sulfates family that may arise by partial replacement or complete substitution with other monovalent and divalent cations. Both K and Pb varieties have been identified in mine drainage precipitates (Chapman et al., 1983; Alpers et al., 1994).

10.5.5. Precipitation of arsenic-bearing minerals

Secondary arsenates form in oxidized zones of tailings and waste rock piles by alteration of arsenopyrite and other minerals. **Scorodite ($Fe^{3+}AsO_4.2H_2O$)** is the most common arsenate mineral, that forms by alteration of arsenopyrite in tailings piles from base metal and gold mines. **Mansfieldite ($AlAsO_4.2H_2O$)**, the aluminum analogue of scorodite, is found in Al-rich environments (Alpers et al., 1994). **Beudantite ($PbFe^{3+}_3(AsO_4)(SO_4)(OH)_6$)** and other arsenic-bearing minerals of the Beudantite group may also be present.

The secondary arsenic-bearing minerals in tailings piles at several gold mining operations have been studied. In particular, Paktunc et al. (1998) found that the secondary arsenic-bearing minerals in the cyanidation tailings pile, deposited from 1988 to 1990, at Ketza River Gold mining operation in Yukon, Canada, are scorodite, an FeCa arsenate, an FeBi arsenate and an arsenic-bearing Fe hydroxide (goethite). The main primary sulfide and arsenide minerals in the ore were pyrrhotite, pyrite and arsenopyrite. In another study, McCreadie et al. (1998) found that the secondary arsenic-bearing minerals in a pile that contained both mill tailings and autoclave tailings from a gold mining operation in the Red Lake area in Ontario, Canada, are ferric sulfarsenate, arsenic-bearing hydronium jarosite, and arsenic-bearing Fe oxides (goethite). The primary sulfide and arsenide minerals in the tailings are pyrrhotite and arsenopyrite. McCreadie et al. (1998) also found that the pore waters in the tailings pile had high concentrations of Fe^{2+}, As^{3+}, Ca and K, which indicates that the secondary arsenic-bearing minerals are unstable and are redissolving. The tailings from the autoclave were deposited between 1991 and 1994.

Reactions of arsenic-bearing minerals in tailings from a uranium mine were determined from a lysimeter test on arsenic-bearing tailings from the Midwest Uranium mine in Saskatchewan (Petruk and Pinard, 1988). The tailings were composed of silicate rock forming minerals; sulfides including pyrite, galena and millerite; oxides including uraninite; arsenides including gersdorffite, rammelsbergite, and nickeline; and calcite. At the end of the test, which was conducted to simulate weathering for the equivalent of 10 years, the calcite was depleted and the Ca had precipitated as Ca sulfates, commonly as a cementing medium. The pyrite was partly decomposed and the individual pyrite grains in some of the massive pyrite particles were delineated (Figure 10.7). The rammelsbergite, nickeline and gersdorffite were partly consumed (Figure 10.8, 10.9) and a Ni-arsenate had precipitated as an armour on the gersdorffite, uraninite, pyrite, and galena particles (Figure 10.10, 10.11, 10.12). Smaller amounts of a Pb hydroxide, Ni sulfate, and a U-Pb silicate had also precipitated as an armour on the particles.

10.5.6. Precipitation of phosphates

Secondary Phosphate minerals found in tailings include **corkite ($PbFe^{3+}_3(PO_4)(SO_4)(OH)_6$)** and other minerals of the corkite group where PO_4^{3-} has substituted for SO_4^{2-} in alunite. The source of phosphate is likely the destruction of primary apatite by sulfuric acid solutions (Stoffregen and Alpers, 1987).

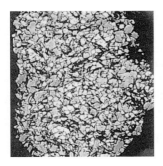

Figure 10.7. Partly oxidized massive pyrite, displaying individual pyrite grains.

Figure 10.8. Rammelsbergite (white) partly replaced by a Ni-arsenate (grey).

Figure 10.9. Gersdorffite (white) partly replaced by a Ni-arsenate (grey)

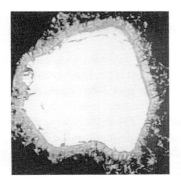

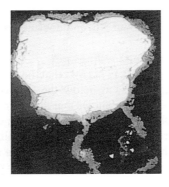

Figure 10.10. Gersdorffite (white) bordered by a Ni-arsenate (grey).

Figure 10.11. Galena (white) bordered by a Ni-arsenate (grey).

Figure 10.12. Uraninite (white) bordered by a Ni-arsenate (grey).

10.5.7. Predicting secondary minerals

A geochemical model, MINTEQA2, is commonly used to predict precipitation of secondary minerals from the pore water solutions. The model calculates the saturation indices of various minerals from the concentrations of the elements in the pore water. MINTEQA2 uses thermodynamic databases to calculate equilibrium reactions and to establish the distribution of the masses of various solid or dissolved species or complexes. The results are compared with actual secondary minerals determined by the mineralogical examination (Allison et al., 1990, 1991; Alpers et al., 1994, vanHuyssteen, 1998a).

10.5.8. Correlating pH, pore water concentrations, rock chemistry and mineralogy in tailings piles

Depth profiles are used to correlate data and to study reactions in tailings piles. Depth profiles, plotted from data published by vanHuyssteen (1998a) for site O are presented in Figure 10.13 as an example of correlations between pH, pore water concentrations, rock chemistry, primary minerals distributions and secondary mineral distributions

The tailings pile at site O was deposited at a gold mine between 1926 and 1947. Deposition was in a raised stack. The water table is ~1.5 m below the tailings surface. The initial tailings had low pyrite and arsenopyrite contents (~5 wt % S and ~0.4 wt % As), but the tailings deposited at the top 2 m of the pile had high pyrite and arsenopyrite contents (~29 wt % S and 3 wt % As).

10.5.8.1. pH in site O tailings pile

The pH in the vadose zone is 2.16 at 30 cm below the surface of the tailings pile and increases to 5.38 at the 150 cm below the tailings surface (top of water table). The pH in the saturated zone (zone below the water table) is about 5.6, and does not change with depth (Figure 10.13).

10.5.8.2. primary minerals in site O tailings pile

The depth profile for pyrite (Figure 10.13) shows that the pyrite content is depleted at tailings surface and increases with depth to 30 cm below the surface where pyrite depletion stops. Similarly the depth profile for arsenopyrite shows that the arsenopyrite content is depleted at the tailings surface and increases with depth to 22.5 cm below the surface. The depletion indicates oxidation of pyrite and arsenopyrite in the top 30 cm of the tailings pile. In contrast the data by Nesbitt and Jambor (1998) for the Waite Amulet tailings pile show depletion of pyrite to the bottom of the vadose zone (1.8 m), and of pyrrhotite a short distance into the saturated zone below the water table (2.4 m).

The depth profile for ankerite shows that the mineral is absent in the top 60 cm of the tailings pile and is present below this level. This suggests that ankerite is a primary mineral that was dissolved in the acidic top portions of the tailings pile where the pH is lower than 3.0

10.5. 8.3. Pore water chemistry in site O tailings pile

The depth profiles for pore water chemistry show high concentrations of S, Fe and As at 30 cm below the tailings surface. The concentrations decrease with depth to 210 cm below surface where they become relatively constant (Figure 10.13). The S depth profile correlates with the pH depth profile, which shows that the pore water is acidic in the vadose zone with the highest acidity at the top of the zone. The high concentrations of Fe and As in the pore waters of the vadose zone show that these elements are dissolving from the pyrite and arsenopyrite respectively, and probably from secondary minerals.

The depth profiles for Cu and Zn concentrations in pore waters show high concentrations at 30 cm below the surface of the tailings pile and decrease with depth to 210 cm. The Ni concentrations in the pore waters continue to decrease to 330 cm below the surface of the tailings pile. This shows intense dissolution of the metals at the low pH of 2.2 at the top of the tailings

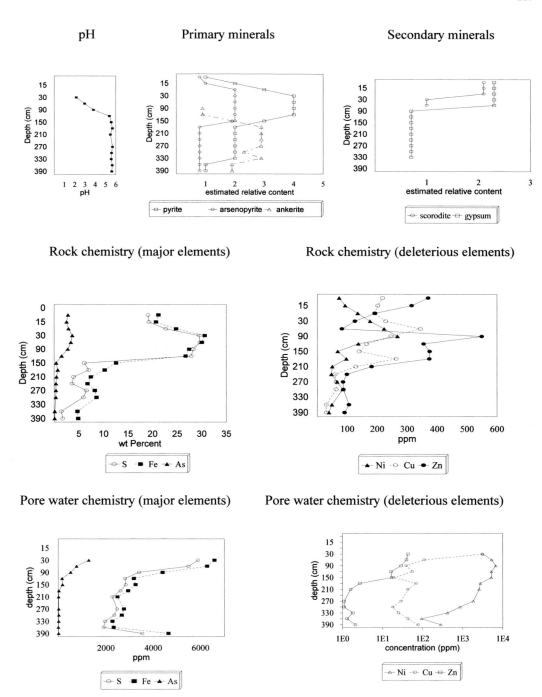

Figure 10.13. Depth profiles showing distributions of pH, primary minerals, secondary minerals, rock chemistry and pore water chemistry in tailings pile from a gold mine at site O.

pile. The dissolution of the metals decreases as the pH increases with depth in the tailings pile.

10.5.8.4. Secondary minerals in site O tailings pile

The presence and quantities of scorodite, gypsum and ankerite were estimated. The depth profile for scorodite shows that the mineral is present only in the top 60 cm of the tailings pile and occurs at high concentrations in the top 22.5 cm (Figure 10.13). This indicates that the arsenopyrite, which dissolved in the vadose zone precipitated as scorodite in the highly acidic parts of the zone. The high concentrations of arsenic in the pore water at 30 cm suggest that some of the scorodite may be dissolving in the acidic portion of the vadose zone.

The depth profile shows a high content of gypsum in the top 60 cm and a low content from 60 to 330 cm. This indicates that the Ca, which undoubtedly dissolved from Ca-bearing carbonates in the top 60 cm of the tailings pile, had combined with SO_4 and formed gypsum. Smaller amounts of gypsum continued to precipitate below 60 cm. The precipitation of gypsum was probably influenced by the neutralizing effects of the dissolved carbonates, and the acidity of the pore water.

10.6. REMEDIATION AND PREVENTION OF ACIDIC DRAINAGE FROM MINE WASTES

Many wastes and tailings piles that were deposited on surface at sites of mining operations discharge acidic waters that contain dissolved metals. The period of intense sulfide oxidation and associated generation of contaminated acidic waters occurs during the first few decades after deposition of mine wastes and tailings, but discharge of sulfide oxidation products continues for many decades or centuries (Blowes et al., 1994). The only available remediation programs at wastes and tailings piles that were deposited on surface are :
  (1) collection and treatment of acidic water discharge,
  (2) reduction of the volume of water and oxygen infiltrating into the piles.
It is noteworthy that revegetation over wastes and tailings piles stabilizes the piles and impoundments with respect to aeolian and water erosion, but does little to prevent sulfide oxidation and the attendant discharge of polluted acidic water.

At new mining operations it is possible to design tailings impoundments that will inhibit sulfide oxidation reactions and minimize drainage of acidic waters. The design of the tailings impoundment must consider:
  • the topography of the proposed tailings site,
  • the nature of the underlying strata as this will determine whether there is downward permeability and a potential pathway for metal rich solutions,
  • the tailings - atmosphere interface,
  • the chemical, mineralogical and physical properties of the tailings,
  • the design height of the tailings pile, and the position of the proposed water table.

Remediation of acidic water drainage from existing waste and tailings piles and from new tailings impoundments is discussed in this section under the following topics:
  • treatment of acidic waters,
  • infiltration controls,

- sulfide oxidation controls (tailings management program)

The discussion draws heavily on the paper by Blowes et al. (1994), which covers remediation and prevention of drainage from tailings impoundments.

10.6.1. Treatment of acidic water

Some of the acidic water in tailings piles is discharged at the toe of the piles, and some filters into the ground water below the tailings. Both streams need to be directed to treatment facilities to neutralize the acidity and to precipitate the metals, before releasing the water into the local drainage system. Several techniques have been proposed for treating the acidic waters. They include:

1. on-site treatment facilities,
2. continuous flow reactors,
3. downstream wetlands.

10.6.1.1. On-site treatment facilities

Facilities are in place at many tailings sites throughout North America to collect the acidic waters, to neutralize the acidity, and to precipitate the dissolved metals. The treatment plants vary from relatively simple to automated computer-controlled facilities. The maintenance costs of these facilities are high because large quantities of lime are required to treat the water. Therefore, there is a strong desire to develop and implement alternative systems. Furthermore, the disposal of sludge residues is an environmental concern, as there is a potential for release of dissolved metals through subsequent leaching.

10.6.1.2. Continuous flow reactors

Dvorak et al. (1991) proposed that drainage water from mine wastes be directed through continuous flow reactors that contain solid phase organic carbon. The reactors would induce reduction of sulfates and precipitation of metal sulfides, as well as increase the pH of the water. The continuous flow reactors must be placed inground to catch the acidic waters, which seeped into the ground below the wastes pile, as well as the acidic waters from the toe of the wastes pile. A variant of the system is to treat the discharge from the reactor with limestone (Kepler and McCleary, 1994). The material at the reactor wall must be sufficiently stable to have a lifespan that is economically competitive, and the material at the centre of the reactor must be sufficiently reactive for the reaction to proceed during the short residence time that the contaminated acidic water is in the reactor. Preliminary experiments indicate that treatment of tailings derived water with a continuous flow sulfate reducing reactor is feasible (Blowes et al., 1994).

10.6.1.3. Downstream wetlands

Wetlands, woodwaste and peat can be used to treat acidic waters, as they reduce the acidity of the water and scavenge dissolved metals from the water (Kleinmann et al., 1991). The main requirement for a downstream wetland treatment system is establishment and maintenance of wetlands that will remain stable over long periods of time, and will survive the metal loadings associated with base- and precious-metal tailings (Blowes and Jambor, 1990; Blowes et al., 1994; Coggans et al., 1994). A major concern with the wetland treatment system is the gradual infilling of the wetland basin and the potential for the remobilisation of contaminants. Kleinmann et al. (1991) noted that attenuation of the dissolved metals results from the bacterially catalysed reduction of sulfate to sulfide and the accompanying precipitation of metal sulfides through reactions of the form:

$$CH_2O + SO_4^2 \Rightarrow HS^- + HCO_3^- + H^+ \tag{6}$$

$$Me^{2+} + HS^- \Rightarrow MeS_{(s)} + H^+ \tag{7}$$

where CH_2O represents a labile source of organic carbon, and MeS represents a sulfide precipitate. These reactions increase the pH of the tailings discharge water, and decrease the concentrations of dissolved metals. In wetlands, organic carbon is supplied and replenished through the growth and degradation of wetland plants. The sulfate and dissolved metals are supplied by the acidic drainage water. The systems may be enhanced by using limestone drains to precondition water prior to discharge to the wetland system, and by focussing on beneficial biological activities (Hedin and Watzlaf, 1994; Kalin and van Everdingen, 1987; Ritcey, 1989; Brown 1991; Machemer and Wildeman, 1992).

10.6.2. Infiltration controls

The most effective method of controlling the movement of tailings pore water is to restrict the entrance of meteoric water, surface water and groundwater into the tailings impoundment. Infiltration of meteoric water can be restricted by surface contouring and by placement of low permeability covers, either of natural geologic (clay) or synthetic material. An optimal cover would provide a barrier to both the infiltration of water and to atmospheric oxygen.

Entry of ground water and surface water can be minimized at new tailings impoundments by appropriate site selection and placement of a low permeability base that may include synthetic liners. Where suitable impoundment locations are not available, or where the tailings are already placed, synthetic liners, cut-off walls and diversion trenching may be considered to minimize entry of ground water and surface water, and to direct the acidic ground water to treatment zones.

10.6.3. Sulfide oxidation controls (Tailings Management Programs)

Sulfide oxidation controls form the basis for tailings management programs. Processes that limit sulfide oxidation are:
1. restricting entrance and diffusion of oxygen into mine wastes,
2. using a thickened slurry with improved moisture retaining capabilities,
3. separation of sulfide and gangue minerals (silicates, carbonates, oxides, etc) for separate disposal,
4. enhance sulfate reduction,
5. precipitates that coat sulfide surfaces,
6. precipitation of heavy metals in reservoir minerals.
7. bactericidal controls,
8. formation of cemented "hardpan" layers,
A combination of these and other techniques may be used in a tailings management program. Items 1, 2 and 3 are currently used.

10.6.3.1. Restricting entrance of oxygen

Sulfide oxidation can be controlled by placing oxygen diffusion barriers between the atmosphere and reactive sulfide tailings to restrict the entrance of oxygen into tailings impoundments. Proposals for diffusion barriers include:

1. disposal of tailings into deep lakes,
2. establishing a water pond behind retention dams,
3. establishing a bog on the tailings surface (Kalin and van Everdingen, 1987),
4. placing oxygen-consuming covers,
5. placing dry covers,

To be effective the oxygen diffusion barriers should be put in place shortly after the tailings deposition has ended, as the rate of sulfide oxidation is greatest in the early stages of tailings weathering.

Disposal of tailings into deep lakes has the following advantages:
- the sulfide minerals are isolated from atmospheric oxygen by a thick water cover,
- oxidation is limited to interaction with dissolved oxygen in overlying water.
- tailings are located at the base of the flow system and are isolated from the effects of erosion and catastrophic dam failure,

Pederson et al. (1993, 1994) studied tailings that were deposited in lakes in northern Canada. They found that the sulfide minerals were stable in deep lake environments, and that concentrations of dissolved metals associated with these tailings were low.

A water pond established above tailings behind retention dams limits oxygen ingress into the tailings by slow diffusion of oxygen through the water, and minimizes sulfide oxidation. It may be necessary to add water continually to maintain the water level above the tailings, because the water may drain due to the coarse-grained nature and high permeability of the tailings. The retention dams, therefore, have to be designed to hold all the water that may accumulate behind them (including flood waters), and to withstand catastrophic dam failure on a long-term basis. Water covers have been successfully applied at six uranium tailings sites in the Elliot Lake region in Canada (Berthelot et al., 1999), and a study of the water above tailings impoundments at the Petaquilla porphyry copper property in Panama showed that the Al and sulfate contents in the water were low enough to meet specifications (Sahami and Riehm, 1999).

A bog on the tailings surface limits oxygen ingress into the tailings and minimizes sulfide oxidation by (1) slow diffusion of oxygen through the overlying water, and (2) consuming oxygen through reactions with organic carbon. As with the water pond, water may need to be added continually to maintain the water level above the tailings. A matter of concern is that the organic material in the bog has the potential of releasing organic acids to the tailings surface. The organic acids have the potential of reducing previously precipitated ferric oxyhydroxide minerals, and releasing the metals that coprecipitated with the ferric oxyhydroxides.

Oxygen-consuming covers that are rich in organic carbon in the form of woodwaste, sewage sludge, and industrial by-products have been proposed as barriers for prevention of oxygen diffusion (Reardon and Poscente, 1984; Blenkinsopp, 1991; Broman et al., 1991; Tassé et al., 1994). The intent is that the organic covers would consume oxygen, and thereby prevent oxygen contact with the underlying sulfide minerals. Reardon and Poscente (1984) calculated that the required mass for long-term prevention of sulfide oxidation may be prohibitive. As with the bog above tailings, there is concern that the organic covers have the potential of releasing high concentrations of organic acids to the tailings surface. The organic acids might reduce previously precipitated ferric oxyhydroxide minerals and release the metals that coprecipitated with them.

Dry covers may be constructed of fine-grained material (clay), or of synthetic low O^{2-} diffusivity materials. To be effective the covers must have long-term integrity, hence it is recommended that they be protected from root penetration, frost action and desiccation. Covers composed of fine-grained materials rely on maintaining saturated or near-saturated conditions several metres above the water table, and possibly eliminating the vadose zone(e.g. zone of sulfide oxidation) (Collin and Rasmuson, 1986; Nicholson et al., 1989, 1991; Yanful et al., 1994). The moisture retaining materials are constructed by placing alternating layers of clay and of silt over the coarser-grained tailings.

A dry cover that consisted of a synthetic layer of 25 cm of Cefill fyash-stabilized concrete, covered by 2 m of till, was installed over a waste-rock pile at the Bersbo site in Sweden (Lundgren and Lindahl, 1991). Such a cover is expensive to construct (Sodermark and Lundgren, 1988), and may be susceptible to cracking after installation, due to desiccation or subsidence (Collin, 1987).

10.6.3.2. Enhanced sulfate reduction
The addition of organic carbon to the tailings would enhance sulfate reduction reactions and reprecipitation of metal sulfides. If an organic carbon-rich zone is below the equilibrium water table, the sulfide minerals that reprecipitate by sulfate reduction reactions will be isolated from atmospheric oxygen by the overlying tailings solids and water column, and will not redissolve. Therefore, Blowes (1990) and Blowes and Ptacek (1992) proposed that organic carbon be added to tailings as they are deposited in the tailings impoundment. It is not known whether this technique has been tested.

10.6.3.3. Thickened slurry with moisture retaining capabilities
Deposition of thickened tailings prevents segregation of grain sizes, thus the poorly sorted tailings have moisture retaining characteristics that result in an extensive tension-saturated zone above the water table (Robinsky et al., 1991; Blowes et al., 1994). The tension-saturated zone limits the zone of rapid oxidation to near the tailings surface (Woyshner and St- Arnaud, 1994; Al et al., 1944). Blowes et al. (1994) reported that the tension-saturated zone in the Kidd Creek tailings, where the thickened tailings deposition technique was used, extended >4 m above the water table.

10.6.3.4. Precipitates that coat sulfide mineral surfaces
Several proposals have been presented for controlling sulfide oxidation by coating sulfide mineral surfaces with an insoluble, non-reactive - coatings that isolate the sulfide minerals from oxidants (e.g. O_2^- and Fe^{3+}) and prevent sulfide oxidation. Huang and Evangelou (1994) proposed ferric phosphate as an armouring phase. They observed decreasing rates of sulfide oxidation in samples amended with PO_4^{3-} compared to control samples. Ahmed (1991) proposed ferric oxyhydroxides as an armouring phase, which seems practical because in field settings

altered sulfide minerals are commonly rimmed by ferric oxyhydroxides. Rybock and Anderson (2000) developed a silica micro encapsulation (SME) technology for encapsulating heavy mineral particles in a silica matrix. The encapsulated particles settle and are effectively isolated from the surrounding environment.

10.6.3.5. precipitation of heavy metals in reservoir minerals.

Pöllman (1998) proposed a method of fixing metals by incorporating them into structures of stable minerals such as ellestadite ($Ca_5(SiO_4,PO_4,SO_4)_3(OH,Cl,F)$), perovskite ($CaTiO_3$), alunite-jarosite ($KAl_3(SO_4)_2(OH)_6$ - $KFe^{3+}_3(SO_4)_2(OH)_6$), ettringite ($Ca_6Al_2(SO_4)_3(OH)_{12}.26H_2O$) and others. The precipitation of some of these reservoir minerals requires a thermal process that is economically prohibitive for disposal of mine tailings. Nevertheless, it may be possible to incorporate the metals into some mineral structures by cementing techniques that could be used in a tailings impoundment. Furthermore, some arsenate minerals that precipitate in tailings impoundments are stable under a wide range of pH conditions, and hence are suitable reservoir minerals. The technique has not been tested on mine tailings, but is interesting enough to warrant investigation. Application of the technique would require a thorough knowledge of the mineralogy and chemistry of the tailings pile, as well as of the stabilities of the reservoir minerals.

10.6.3.6. Bactericidal controls

Bactericidal controls involve inhibiting naturally occurring sulfide oxidizing bacteria by applying bactericides either directly to the tailings surface or mixing them with the tailings in the impoundment (Erickson and Ladwig, 1985; Lortie et al., 1999; Stichbury et al., 1995). In the absence of sulfide oxidation bacteria the sulfide-mineral oxidation decreases as the pH decreases, and the concentrations of dissolved metals remain low. The bactericides currently available require continual reapplications, which suggests that bactericides are suited for short term disposal sites, such as temporary storage facilities during impound construction.

10.6.3.7. Cemented "hardpan" layers

Cemented "hardpan" layers inhibit the diffusion of gases into and out of tailings impoundments. Blowes et al. (1991) proposed enhancing the formation of "hardpan" layers by selective layering of tailings during the late stages of deposition, and Ahmed (1994) proposed the addition of $FeSO_4$ solutions to enhance the formation of cemented layers.

10.6.3.8. Separating sulfide minerals from gangue minerals

Separating the sulfide minerals from the gangue minerals would produce two tailings streams, high sulfur and low sulfur tailings. The high sulfur tailings must be handled in a manner that will prevent contact with atmospheric oxygen. In contrast the low sulfur tailings will be relatively inert and can be readily disposed of, as they will not produce acidic water drainage. The approach of separating sulfide and gangue minerals is practical only for ores that have low sulfide contents (e.g. porphyry copper ores). It is not practical for high sulfide ores, such as base metal ores that contain up to 60 % pyrite.

10.7. SUMMARY

Acidic water is generated in tailings and waste rock piles by oxidation of sulfides, particularly, pyrite and pyrrhotite. The main conditions for oxidation to proceed are moisutre and a supply of oxygen. Pyrite oxidizes slowly at high pH values to goethite and elemental sulfur. Neutrophillic thiobacilli bacteria oxidize the elemental sulfur at a pH of 6 to 8.5 to produce sulfuric acid. The sulfuric acid reduces the pH of the pore water. When the pH is reduced to <4.5 acidophilic bacteria, such as thiobacillus ferrooxidans, oxidize the pyrite to Fe^{2+}, elemental sulfur and sulfates, and generate acid.

Reactions involving dissolution of carbonate minerals neutralize the acidic water and maintain the pH at near neutral conditions. Hence, pyritic tailings amended with fine-grained limestone do not produce acidic water, but the volume of limestone must be at least 12 times the volume of pyrite, before the tailings can be considered as having no acid generating potential.

Following depletion of calcite the pore water pH drops abruptly to 4.8 and siderite begins to dissolve. When the siderite is depleted the pH drops to the next buffer viz $Al(OH)_3$. When the $Al(OH)_3$ is consumed the pH drops to <3.5 and the sulfides such as chalcopyrite, sphalerite, arsenopyrite, etc. dissolve, with the aid of bacterial activity, and release deleterious elements (e.g. Cu, Zn, Pb, Cd, As, Sb, Se, U, Ni, Co, Hg, etc.) into the pore water. The neutralizing effects of aluminosilicates become significant at this stage and the pore water is enriched in Al and Mg.

Remediation of acidic drainage from tailings and waste rock piles deposited on surface is limited to:
(1)   treating acidic waters to neutralize the water and precipitate the metals by:
         (a) on-site treatment facilities (widely used).
         (b) continuous flow reactors
         (c) downstream wetlands.
(2)   minimizing or eliminating infiltration of water and oxygen into the piles by placing low permeability natural covers (clay) and impermeable synthetic covers over the piles.

Deposition of new tailings and waste rock piles must be conducted according to specifications and conditions outlined by local Tailings Management Programs for controlling sulfide oxidation. The options are:
(1) disposing tailings into deep lakes (preferable because sulfide minerals are stable in deep lake environments).
(2) preparing tailings impoundments. Tailings impoundments require:
         (a) selection of an appropriate site. Low permeability liners, that consist of natural (e.g. clay) and/or synthetic materials, are commonly placed in the impoundment to prevent leakage into the ground water.
         (b) preparation of the tailings surface to restrict entrance of oxygen. Some oxygen diffusion barriers between the atmosphere and reactive sulfide tailings are:
              (i) water cover on tailings behind retention dams,
              (ii) bog on the tailings surface,
              (iii) an oxygen consuming cover that is rich in organic carbon,
              (iv) dry covers on tailings piles.
         (c) treatment of the material in the tailings impoundment by:
              (i) mixing carbon into the tailings to enhance sulfate reduction,
              (ii) thickening the slurry to raise the watertable in the impoundment,

(iii) encapsulating reactive sulfide minerals with non-reactive coatings (e.g. goethite, phosphates, silicates) (experimental),

(iv) fixing deleterious metals and sulfates into the structure of non-reactive phases (experimental),

(v) enhancing formation of hardpan layers,

(vi) using bacteria that will inhibit sulfide oxidation (short term).

(3) Separating reactive sulfide minerals from stable gangue minerals and storing the products separately. The pile of stable gangue minerals will not require further treatment, whereas the pile of reactive tailings would have to be treated . This approach is applicable to ores that contain small amounts of sulfides.

REFERENCES

R.N. Adair, Stratigraphy, Structure, and Geochemistry of the Halfmile Lake Massive-sulfide Deposit, New Brunswick. *Explor. Mining Geol.,* 1, 2 (1992) 151-166.

S.M. Ahmed, Electrochemical and Surface Chemical Methods for the Prevention of Atmospheric Oxidation of Sulphide Tailings and Acid Generation, *Proceedings 2nd Internat. Conf. Abatement Acidic Drainage 2,* MEND Secretariat, Ottawa, Canada (1991) 305-319.

S.M. Ahmed, Surface Chemical Methods for the Prevention of the Atmospheric Oxidation of Sulphide Tailings and Acid Generation, *Proceedings Internat. Land Reclamation Mine Drainage Conf. and 3rd Internat. Conf. Abatement Acidic Drainage 2*, U.S. Dept. Interior, Bureau of Mines, Special Pub. SP 06A-94 (1994) 57-66.

T.A. Al, D.W. Blowes and J.L. Jambor, The Pore Water Geochemistry of the Cu-Zn Mine Tailings at Kidd Creek Near Timmins, Ontario, Canada, *Internat. Land Reclamation Mine Drainage Conf. and 3rd Internat. Conf. Abatement Acidic Drainage 2*, U.S. Dept. Interior, Bureau of Mines, Special Pub. SP 06A-94 (1994) 208-217.

L. Alexander, and H.P. Klug, Basic Aspects of X-ray Absorption in Quantitative Diffraction Analysis of Powder Mixtures, *Anal. Chem.,* 20 (1948) 886.

J.D. Allison, D.S. Brown and K.J. Nova-Gradac, MINTEQA2/PRODEFA2, A Geochemical Assessment Model for Environmental Systems: Version 3 User's Manual, Environmental Research Laboratory, Office of Research and Development, U.S. Environmental Protection Agency, Athens, Georgia (1990).

J.D. Allison, D.S. Brown and K.J. Nova-Gradac, MINTEQA2/PRODEFA2, A Geochemical Assessment Model for Environmental Systems: Version 3 User's Manual, Environmental Research Laboratory, Office of Research and Development, U.S. Environmental Protection Agency, Athens, Georgia, Report EPA/600/3-91/021 (1991).

C.N. Alpers, D.W. Blowes, D.K. Nordstrom and J.L. Jambor, Secondary Minerals and Acid Mine-water Chemistry, in *Environmental Geochemistry of Sulfide Mine-Wastes*, eds. J.L. Jambor and D.W. Blowes, Mineral. Assoc. Can., Short Course Vol. 22 (1994) 247-270.

C.N. Alpers and D.K. Nordstrom, Geochemical Evolution of Extremely Acid Mine Waters at Iron Mountain, California: Are There any Lower Limits to pH?, *Proceedings 2nd Internat. Conf., Abatement Acid Drainage 2*, MEND Secretariat, Ottawa, Canada (1991) 321-342.

P.R.A. Andrews, Summary Report No.13: Major Abrasives - Garnet, Industrial Diamond, Silicon Carbide, and Fused Alumina, Mineral Sciences Laboratories, CANMET, Dept. Energy, Mines and Resources, Canada, Division Report MSL 91- 81(R) (1991) 112p.

P.R.A. Andrews, Summary Report No.18: Wollastonite, Mineral Sciences Laboratories, CANMET, Dept. Energy, Mines and Resources, Canada, Division Report MSL 93-10(R) (1993) 28p.

P.R.A. Andrews and J.A. Soles, A Beneficiation Study of Talc from Canada Talc Industries, Madoc, Mineral Sciences Laboratories, CANMET, Dept. Energy, Mines and Resources, Canada, Division Report MSL 85-6(CR) (1985).

J.P.N. Badham, Further Data on the Formation of Ores at Rio Tinto, Spain, *Trans. Instn. Min. Metall.*, (Sec. B: Appl. Earth Sci.) 91 (1982) B26-B32.

G. Barbery, Mineral Liberation: Measurement, Simulation, and Practical Use in Mineral Processing. Les Editions GB, Quebec (1991) 351p.

R.W. Bartlett and K.A. Prisbrey, Oxygen Diffusion into Wet Ore Heaps Impeded by Water Vapour Upflow, in *Global Exploitation of Heap Leachable Gold Deposits*, eds. .D.M. Hausen, W. Petruk and R.D. Hagni, Mineral, Metals and Materials Society, TMS, Warrendale, PA. (1997) 85-94.

R.W. Bartlett, Economic Criteria for Choosing Bioheap Treatment of Mixed Oxide/Refractory Gold Ore, in *Global Exploitation of Heap Leachable Gold Deposits*, eds. D.M. Hausen, W. Petruk and R.D. Hagni, Mineral, Metals and Materials Society, TMS, Warrendale, PA (1997) 95-103.

P.B. Barton, Jr., and P.M. Bethke, Chalcopyrite Disease in Sphalerite: Pathology and Epidemiology, *Am. Mineral.*, 72 (1987) 451-467.

G.M. Bernard, Andacollo Gold Production - Ahead of Schedule and Under Budget, *Mining Engineering*, 48, 8 (1996) 42-47.

D. Berthelot, M. Haggis, R. Payne, D. McClarity and M. Courtin, Application of Water Covers, Remote Monitoring and Data Management Systems to Environmental Management at Uranium Tailings Sites in the Serpent River Watershed, *CIM Bull.*, 92, 1033 (1999) 70-77.

M. Bezuidenhout, J.S.J.vanDeventer, and D.W. Moolman, Identification of Plant Disturbances in Froth Flotation Using Computer Vision, *CIM Bull.* (2000) In press.

M. Bhargava and A.B. Pal, Anatomy of a Porphyry Copper Deposit - Malanjkhand, Madhya Pradesh, *Geol. Soc. India*, 53 (1999) 675-691.

A.C.T. Bigg, The Nanisivik Mill, *Proceedings, 12th Annual Meeting, Canadian Mineral Processors*, compiled by R.W. Bruce, CIM, Mineral Processors Division, Ottawa, Canada, paper 3 (1980) 16p.

J.M. Bigham, Mineralogy of Ochre Deposits Formed by Sulfide Oxidation, in *Environmental Geochemistry of Sulfide Mine-Wastes*, eds. J.L. Jambor and D.W. Blowes , Mineral. Assoc. Can., Short Course Vol. 22, (1994) 103-132.

S.A. Blenkinsopp, The Use of Biofilm Bacteria to Exclude Oxygen From Acidogenic Mine Tailings, in *Proceedings 2nd Internat. Conf. Abatement Acidic Drainage 1*, MEND Secretariat, Ottawa, Canada (1991) 369-377.

A.M. Blount, Detection and Quantification of Asbestos and Other Trace Minerals in Powdered Industrial-mineral Samples, in *Process Mineralogy IX,* eds, W. Petruk, R.D. Hagni, S. Pignolet-Brandom, and D.M. Hausen, TMS, Warrendale, PA (1990) 557-570.

A.M. Blount and B.H. McHugh, Exploration for Buried Talc Bodies Through Analyses of Residual Soils, in *Process Mineralogy VI*, ed. R.D. Hagni, TMS, Warrendale, PA (1986) 291-314.

D.W. Blowes, The Geochemistry, Hydrogeology and Mineralogy of Decommissioned Sulfide Tailings: A Comparative Study, Ph.D. thesis, University of Waterloo, Waterloo, Ontario, Canada (1990).

D.W. Blowes and J.L. Jambor, The Pore Water Geochemistry and Mineralogy of the Vadose Zone of Sulfide Tailings, Waite Amulet, Québec, Canada, *Applied Geochem.*, 5 (1990) 327-346.

D.W. Blowes, J.L. Jambor, E.C. Appleyard, E.J. Reardon and J.A. Cherry, Temporal Observations on the Geochemistry and Mineralogy of a Sulfide-rich Mine-tailings Impoundment, Heath Steele Mines, New Brunswick, *Explor. Mining Geol.*, 1 (1992) 251-264.

D.W. Blowes and C.J. Ptacek, Treatment of Mine Tailings, US Patent (1992).

D.W. Blowes and C.J. Ptacek, Acid-neutralization Mechanisms in Inactive Mine Tailings, in *Environmental Geochemistry of Sulfide Mine-Wastes*, ed. J.L. Jambor and D.W. Blowes, Mineral. Assoc. Can., Short Course Vol. 22, (1994) 271-292.

D.W. Blowes, C.J. Ptacek and J.L. Jambor, Remediation and Prevention of Low-quality Drainage from Tailings Impoundments, in *Environmental Geochemistry of Sulfide Mine-Wastes*, eds. J.L. Jambor and D.W. Blowes, Mineral. Assoc. Can., Short Course Vol. 22 (1994) 365-379.

D.W. Blowes, E.J. Reardon, J.L. Jambor and J.A. Cherry, The Formation and Potential Importance of Cemented Layers in Inactive Sulfide Mine Tailings, *Geochim. Cosmochim. Acta,* 55 (1991) 965-978.

G. Bonifazi, P. Massacci, and A. Meloni, DFSM (Digital Surface Modeling): A 3D Froth Surface Rendering and Analysis to Characterize Flotation Processes, Univ. of Rome, in Abstracts, 2000 SME Annual Meeting (2000) 41.

230

G. Bonifazi, F. La Marca and P. Massacci, Bulk Particle Characterization by Image Analysis, in *Proceedings XX IMPC*, 1 (1997a) 327-342.

G. Bonifazi, P. Massacci, G. Patrizi and G. Zannoni, Colour Classification for Glass Recycling, in *Proceedings XX IMPC*, 5 (1997b) 239-252.

R. S. Boorman and D. Abbott, Indium in Co-existing Minerals from the Mount Pleasant Tin Deposit, *Can. Mineral., 9* (1967) 166-179.

R.S. Boorman and D.M. Watson, Chemical Processes in Abandoned Sulfide Tailings Dumps and Environmental Implications for Northeastern New Brunswick, *CIM Bull.* 69, 772 (1976) 86-96.

D.R. Boyle, Geochemistry and Genesis of the Murray Brook Precious Metal Gossan Deposit, Bathurst Mining Camp, New Brunswick, *Exploration and Mining Geol.*, 4, 4 (1995) 341-363.

R.W. Boyle, The Geochemistry of Gold and its Deposits, Geol. Survey of Canada, *Bull. 280* (1979).

A. Branham and B. Arkell, The Mike Gold-Copper Deposit, Carlin Trend Nevada, in *Process Mineralogy XIII* ed. D. Hagni, TMS, Warrendale, PA (1995) 203-211.

T.B. Braun and P.L. LeHoux, Gold Leaching and Recovery Operations at Barneys Canyon Mine, *Proceedings Milton E. Wadsworth (IV) International Symposium on Hydrometallurgy,* eds. J.B. Hisky and G.W. Warren, Littleton, CO: SEM (1993) 521-529.

P.G. Broman, P. Haglund and E. Mattsson, Use of Sludge for Sealing Purposes in Dry Cover - Development and Field Experiences, *Proceedings 2nd Internat. Conf. Abatement Acidic Drainage 1,* MEND Secretariat, Ottawa, Canada (1991) 515-528.

A. Brown, Proposal for the Mitigation of Acid Leaching from Tailings Using a Cover of Muskeg Peat, *Proceedings 2nd Internat. Conf. Abatement Acidic Drainage 2,* MEND Secretariat, Ottawa, Canada (1991) 517-527.

R.C. Burtan, Silicosis, an Ancient Malady in a Modern Setting, AIME/SME, Littleton, CO, Preprint 84-81 (1984) 5p.

L.J. Cabri and J.L. Campbell, The Proton Microprobe in Ore Mineralogy (micro-PIXE technique), in *Modern Approaches to Ore and Environmental Mineralogy*, eds. L.J. Cabri, and D.J. Vaughan, Mineral. Assoc. Can., Short Course Vol. 27 (1998) 181-198.

L.J. Cabri, J.L. Campbell, J.H.J. Laflamme, R.G. Leigh, J.A. Maxwell, and J.D. Scott, Proton-microprobe Analysis of Trace Elements in Sulfides from some Massive-sulfide Deposits, *Can. Mineral. 23,2* (1985) 133-148.

L.J. Cabri, S.L. Chryssoulis, J.P.R. DeVilliers, J.H.G. Laflamme, and R. Buseck, The Nature of

Invisible Gold in Arsenopyrite, *Can. Mineral. 27, 3* (1989) 353-362.

L.J.Cabri and S.L. Chryssoulis, Genesis of the Olympias Carbonate-hosted Pb-Zn (Au,Ag) Sulfide Ore Deposit, Eastern Chalkidiki Peninsula, Northern Greece - A Discussion, *Econ. Geol.*, 85 (1990) 651-652.

L.J. Cabri and G. McMahon, SIMS Analysis of Sulfide Minerals for Pt and Au: Methodology and Relative Sensitivity Factors (RSF), *Can. Mineral.,* 33 (1995) 349-359.

L.J. Cabri, W. Petruk, J.H.J. Laflamme, and J. Robitaille, Quantitative Mineralogical Balance for Major and Trace Elements in Samples from Agnico-Eagle Mines Limited, Quebec, Canada, in *Analytical Technology in the Mineral Industries*, eds. L.J. Cabri, TMS, Warrendale, PA (1999) 177-192.

L. Carlson and U. Schwertmann, Natural Ferrihydrites in Surface Deposits from Finland and their Association with Silica, *Geochim. Comochim. Acta*, 45 (1981) 421-429.

J.M. Carr and A.J. Reed, Afton: A Supergene Copper Deposit, in *Porphyry Deposits of the Canadian Cordillera*, ed. A. Sutherland Brown, CIM, Special Vol. 15 (1976) 376-387.

D.J.T. Carson, J.L. Jambor, P.L. Ogryzlo and T.A. Richards, Bell Copper: Geology, Geochemistry and Genesis of a Supergene-enriched, Biotitized Porphyry Copper Deposit with a Superimposed Phyllic Zone, in *Porphyry Deposits of the Canadian Cordillera*, ed. A. Sutherland Brown, CIM, Special Vol. 15 (1976) 245-263.

D.J.T. Carson and J.L. Jambor, Morrison: Geology and Evolution of a Bisected Annular Porphyry Copper Deposit, in *Porphyry Deposits of the Canadian Cordillera*, ed. A. Sutherland Brown, CIM, Special Vol. 15 (1976) 264-273.

S. Castro, J. Goldfarb and J. Laskowski, Sulphidizing Reactions in the Flotation of Oxidized Copper Minerals, 1. Chemical Factors in the Sulphidization of Copper Oxide, *Int. Jour. Mineral Processing, 1* (1974) 141-149.

E.C.T. Chao, J.R, Chen, J.A. Minkin and J.M. Back, Synchrotron and Micro-Optical Studies of the Occurrence of Gold in Carlin-Type Ore Samples: Implications for Processing and Recovery, *Process Mineralogy VII*, eds. H, Vassilou, D.M. Hausen and D.J.T. Carson, TMS, Warrendale, PA (1987)143-152.

B.M. Chapman, D.R. Jones and R.F. Jung, Processes Controlling Metal Ion Attenuation in Acid Mine Drainage Streams, *Geochim. Comochim. Acta*, 47 (1983) 1957-1973.

T.T. Chen, Colloform and Framboidal Pyrite from the Caribou Deposit New Brunswick, *Canadian Mineralogist*, 16 (1978) 9-15.

T.T.Chen and W. Petruk, Mineralogy and Characteristics that Affect Recoveries of Metals and Trace Elements from the Ore at Heath Steele Mines, New Brunswick, *CIM Bull.*, 73, 823

232

(1980) 167-179.

T.T.Chen and J.E. Dutrizac, Practical Mineralogical Techniques for the Characterization of Hydrometallurgical Products, in *Process Mineralogy IX*. eds.. Petruk, R.D. Hagni, S. Pignolet-Brandom and D.M. Hausen, TMS, Warrendale, PA (1990) 289-309.

T. Chong, B. Marchand, Z. Nikic, Flotation of a Copper-Gold Oxide-Sulphide Ore, *Proceedings, Canadian Mineral Processors,* compiled by L. Tyreman and K. Meyer, CIM, Mineral Processors Division, Ottawa, Canada (1991) paper 12.

S.L. Chryssoulis, Detection and Quantification of 'Invisible' Gold by Microprobe Techniques, *Gold 90*, ed. D.M. Hausen, AIME/SME, New York, NY (1990).

S.L. Chryssoulis and L.J. Cabri, Significance of Gold Mineralogical Balances in Mineral Processing, *Trans. Instn. Min. Metall.*, Sec. C: Mineral Process. Extr. Metall. 99, T.T. (1990) C1-C10.

S.L. Chryssoulis and N. J. Cook, Determination of Invisible Gold Within Sulphide Minerals in Flotation Tailings from Trout Lake Mine, Manitoba, Surface, *Science Western, Report SSW/PM./88/11* (1988) 42p.

S.L Chryssoulis and J. Kim, Surface Characterization of Pyrite and Sphalerite from the Brunswick Grinding, Classification and Cu-Pb Flotation Circuits, Company report, Noranda Technology Centre (1994).

S.L. Chryssoulis, J.Y. Kim and K. Stowe, LIMS Study of Variables Affecting Sphalerite Flotation, *Proceedings, 26th Annual Meeting, Canadian Mineral Processors*, compiled by L. Buckingham and E. Robles, CIM, Mineral Processors Division, Ottawa, Canada, Paper 28 (1994) 16p.

S.L. Chryssoulis, K.J. Stowe and F. Reich, Characterization of Composition of Mineral Surfaces by Laser-probe Microanalysis, *Trans. Instn. Min. Metall.*, Sect. C. Mineral Process. Extr. Metall, 101 (1992) C1-C6.

S.L. Chryssoulis, L.J. Surges and R.S. Salter, Silver Mineralogy at Brunswick Mining and Smelting and its Implications for Enhanced Recovery, in *Complex Sulphides, Processing of Ores, Concentrates and By-products*, eds. A.D. Zunkel, R.S. Boorman, A.E. Morris and R.J. Wesley, AIME-TMS, Warrendale , PA (1985) 815-830.

S.L. Chryssoulis and Rong Y.Wan, Mineralogic Evaluation of Unleached Gold in Biooxidized Leach Residues, in *Global Exploitation of Heap Leacheable Gold Deposits*, eds. D.M. Hausen, W. Petruk, R.D. Hagni, TMS, Warrendale, PA (1997) 119-127.

S.L.Chryssoulis, C. Weisener, C. Wong, and J. Kim, Mineral Surface Study of the Brunswick Zinc Circuit by TOF-LIMS, Amtel report 95/17(CR) (1995) 52p.

S.L. Chryssoulis and A.H. Winckers, Effect of Lead Nitrate on the Cyanidation of David Bell Ore, in *Proceedings, 28th Annual Meeting, Canadian Mineral Processors*, compiled by M. Mular, CIM, Mineral Processors Division, Ottawa, Canada (1996)127-149.

G. Chung, To Improve Iron Ore Recovery while Meeting Varied Customer Demands, Canada/Newfoundland Mineral Development Agreement, DSS, SSC File No. 006SQ.23440-1-9124 (1992).

J.E. Clemson, Delimiting Heap Leachable Zones in the Black Pine Gold Deposit, Idaho, *Process Mineralogy VIII*, eds .J.T. Carson and H. Vassilou, TMS, Warrendale, PA (1988) 57-79.

C.J. Coggans, D.W. Blowes, W.D. Robertson and J.L. Jambor, The Hydrogeochemical Evolution of a Nickel Mine Tailings Impoundment, Copper Cliff, Ontario, *Reviews in Econ. Geol.* (1994).

M. Collin, Mathematical Modelling of Water and Oxygen Transport in Layered Soil Covers for Deposits of Pyrite Mine Tailings, *Licentiate Treatise*, Royal Institute of Technology, Stockholm, Sweden (1987).

M. Collin and A. Rasmuson, Distribution and Flow of Water in Unsaturated Layered Cover Materials for Waste Rock, National Swedish Environmental Protection Board, Report 3088 (1986).

R.K. Collings and P.R.A. Andrews, Summary Report No. 4: Silica, Mineral Sciences Laboratories, CANMET, Dept. Energy, Mines and Resources, Canada, Report 89-1E (1989) 127p.

R.K. Collings and P.R.A. Andrews, Summary Report No. 8: Talc and Pyrophyllite, Mineral Sciences Laboratories, CANMET, Dept. Energy, Mines and Resources, Canada, Division Report MSL 90-18(R) (1990) 115p.

N.J. Cook and S.L. Chryssoulis, Concentration of 'Invisible gold' in the Common Sulfides, *Can. Mineral.*, 28 (1990) 1-16.

D.R. Cousens, R. Rasch and C.G., Ryan, Detection Limits and Accuracy of the Electron and Proton Microprobe, *Micron*, 28 (1970) 231-239.

J.R. Craig, M.S. Najar and A.M.Robin, Characterization of Coke Gasification Slags, in *Process Mineralogy IX*. eds. W. Petruk, R.D. Hagni, S. Pignolet-Brandom and D.M. Hausen, TMS, Warrendale, PA (1990) 473-483.

A.J. Criddle, Ore Microscopy and Photometry (1890-1998), in *Modern Approaches to Ore and Environmental Mineralogy*, eds. L.J.Cabri and D.J. Vaughan, Mineral. Assoc. Can., Short Course Vol. 27 (1998) 1-74.

J. De Cuyper, Flotation of Copper Oxide Ores, *Erzmetal*, Band XXX (1977).

A. Delesse, Procédé Mécanique pour Déterminer la Composition des Roches, *Annales des Mines,* 13, 4 (1848) 379-388.

D.H. Dvorak, R.S. Hedin, H.M. Edenborn and S.L. Gustafson, Treatment of Metal-contaminated Water Using Bacterial Sulfate-reduction: Results from Pilot Plant Reactors, in *Proceddings 2ⁿᵈ Internat. Conf. Abatement Acidic Drainage 1,* MEND Secretariat, Ottawa, Canada (1991) 588-598.

M.S. Enders, M.W. Bartlett, G.C. Griffin, J.E. Volberding and D.P. Young, Discovery of the McDonald Gold Deposit, *Mining Engineering,* 47,10 (1995) 916-921.

P.E. Epstein, The medical aspect update: working with asbestos - recent studies and risks, AIME/SME, preprint No. 84-135 (1984) 2p.

P.M. Erickson and J. Ladwig, Control of Acid Formation by Inhibition of Bacteria and by Coating Pyrite Surfaces, West Virgina Dept. of Natural Resources (1985) 68p.

R.G. Fandrich, C.L. Schneider and S.L. Gay, Two Stereological Correction Methods: Allocation Method and Kernel Transformation Method, *Minerals Engineering,* 11,8 (1998) 707-715.

D.S. Fanning, M.C. Rabenhorst and J.M. Bigham, Colors of Acid Sulfate Soils, in *Soil Color,* ed. J.M. Bigham and E.J. Ciolkosz, Soil Sci. Soc. Am., Madison, Wisconsin (1993) 91-108.

D.G. Feasby and G.A. Tremblay, Role of Mineral Processing in Reducing Environmental Liability of Mine Wastes, in *Proceedings, 27ᵗʰ Annual Meeting, Canadian Mineral Processors,* complied by L. Duval and S. Laplante, CIM, Mineral Processors Division, Ottawa, Canada (1995) 217-234.

D.G. Feasby, G.A. Tremblay and C.J. Weatherell, A Decade of Technology Improvement to the Challenge of Acid Drainage - A Canadian Perspective, in *Proceedings Fourth Internat. Conf. On Acid Rock Drainage, 1,* MEND, CANMET, Natural Resources Canada, Ottawa (1997) i-ix.

J.A. Finch and W. Petruk, Testing a Solution to King Liberation Model, *Int. Journ. Mineral Processing,* 12 (1984) 305-311.

M.E. Fleet, S.L. Chryssoulis, P.J. MacLean, R. Davidson and C.G. Weisener, Arsenian Pyrite from Gold Deposits: Au and As Distribution Investigated by SIMS and EMP, and Color Staining and Surface Oxidation by XPS and LIMS, *Can. Mineral.,* 31 (1993) 1-17.

M. Fleischer and J.A. Mandarino, Glossary of Mineral Species, Glossary - Mineralogical Record, Tucson, Arizona (1995).

C.A. Fleming, Thirty Years of Turbulent Change in the Gold Industry, in *Proceedings, 30ᵗʰ annual meeting, Canadian Mineral Processors,* complied by C. Edwards, CIM, Mineral Processors Division, Ottawa, Canada (1998) 49-82.

Food and Drug Administration, Asbestos Particles in Food and Drugs, Federal Register (1973) 27076-27081.

J..M. Franklin, J.W. Lydon and D.F. Sangster, Massive Sulphide Deposits, *Econ. Geol* 75th Anniversary Volume (1981) 485-627.

E. Fregeau-Wu, S. Pignolet-Brandom and I. Iwasaki, In-situ Grain Size Determination of Slow-cooled Steelmaking Slags with Implications to Phosphorous Removal for Recycle, in *Process Mineralogy IX*, eds. W. Petruk, R.D. Hagni, S. Pignolet-Brandom, and D.M. Hausen TMS, Warrendale, PA (1990) 429-439.

O.C. Gaspar, Paragenesis of the Neves-Corvo Volcanogenic Massive Sulphides, *Comun. Serv. Geol. Portugal,* 77 (1991) 27-52.

C. Gasparrini, The Mineralogy of Gold and its Significance in Metal Extraction, *CIM, Bull.*, 76, 851 (1983)144-153.

G. Gateau and A. Broussaud, New Approaches to the Interpretation of One and Two Dimensional Measurements of Mineral Liberation, *Acta Stereologica*, 5, 2 (1986) 397-402.

M.I. Gorodetskii, N.N. Yaschenko, N.E. Plaksa, L.I. Mekler and A.M. Goldman, Flotation Recovery of Porphyry Copper Ores at the Balkhash Concentrator, Tenth International Mineral Processing Congress, Institution of Mining and Metallurgy (1973) 689-705.

W.D. Gould, The Role of Microbiology in the Prevention and Treatment of Acid Mine Drainage, in *Workshop, Mining and Environment*, eds. R.C. Villas Boas, R. Hargreaves and J.P. Barbosa, CETEM, Rio de Janeiro, Brazil (1997) 94-110.

W.D. Gould, G. Béchard and L. Lortie, The Nature and Role of Microorganisms in the Tailings Environment, in *Environmental Geochemistry of Sulfide Mine-Wastes*, eds. L. Jambor and D.W. Blowes, Mineral. Assoc. Can., Short Course Vol. 22 (1994) 185-199.

R. Grant, C. Bazin and M. Cooper, Optimizing Grind Using Size by Size Recovery Analysis at Brunswick Mining and Smelting Corporation, in *Evaluation and Optimization of Metallurgical Performance*, eds. D.Malhotra, R.R. Klimpel and A.L. Mular, AIME/SME, Littleton, CO (1991) 329-336.

R. Green and G. B. Mellon, Geology of the Chinchaga River and Clear Hills (north half) Map Areas, Alberta, Res. Council Alta., Prelim. Report 62-8 (1962).

G.A. Gross, General Geology of Iron Deposits in Canada, Vol. I, General Geology and Evaluation of Iron Deposits, Geological Survey of Canada, Econ. Geol. Report, 22 (1965).

G.A. Gross, General Geology of Iron Deposits in Canada, Vol. III, Iron Ranges of the Labrador Geosyncline, Geological Survey of Canada, Econ. Geol. Report, 22 (1968).

J. Guha, A. Gauthier, M. Vallee, J. Descarreaux, and F. Lange-Brard, Gold Mineralization Patterns, at the Doyon Mine (Silberstack), Bousquet, Quebec, in *Geology of Canadian Gold Deposits*, eds. R.W. Hodder and W. Petruk, CIM, Special Volume 24 (1982) 50-57.

R.D. Hagni, Cathodoluminescence Microscopy Applied to Mineral Exploration and Beneficiation, in *Applied mineralogy*, eds., W.C. Park, D.M. Hausen and R.D. Hagni, TMS, Warrendale, PA (1985) 41-66.

R.D. Hagni, Industrial Applications of Cathodoluminescence Microscopy, in *Applied Mineralogy VI*, ed., R.D. Hagni, TMS, Warrendale, PA (1986) 37-52.

K.B. Hall, Homestake Carbon-in-pulp Process, Am. Mining Congress, Las Vegas, Nev. USA (1974) 16p.

A. Hamilton, Geology of the Stratmat Boundary and Heath Steele N-5 Zones, Bathurst Camp, Northern New Brunswick, *Explor. Mining Geol.* 1, 2 (1992) 135-135.

I. Härkönen, K.K. Kojonen and B. Johanson, The Early Proterozoic Suurikuusikko Refractory Gold Deposit, Kittilä, Western Finnish Lapland, in *Proceedings 5th Biennial SGA meeting & 10th IAGOD meeting*, London, UK, ed. C.J. Stanley et. Al., A.A. Balkema Publishers, Rotterdam, Netherlands (1999) 159-162.

M. Harley and E.G. Charlesworth, Structural Development and Controls to Epigenetic, Mesothermal Gold Mineralization in the Sabie-Pilgrim's Rest Goldfields, Eastern Transvaal, South Africa, *Explor. Mining Geol.*, 3, 3 (1994) 231-246.

D.C. Harris, The Mineralogy and Geochemistry of the Hemlo Gold Deposit, Ontario, Geol. Survey of Canada, Report 38 (1989).

D.M. Hausen, Quantitative Measurement of Wallrock Alteration in the Exploration of Buried Mineral Deposits, *Trans. Soc. of Mining Engineers*, 266 (1979) 1853-1860.

D.M. Hausen, Process Mineralogy of Auriferous Pyritic Ores at Carlin, Nevada, in *Process Mineralogy*, ed. D.M. Hausen and W.C. Park, TMS, Warrendale, PA (1981) 271-289.

D.M. Hausen, Process Mineralogy of Auriferous Pyritic Ores at Carlin, Nevada, *Geol. Soc. S.Africa.*, Spec. Publ. 7 (1983) 261-269.

D.M.Hausen, Process Mineralogy of Selected Refractory Carlin-Type Gold Ores, *CIM Bull.*, Sept. (1985) 83-94.

D.M. Hausen, J.W. Ahlrichs, W. Mueller and W.C. Park, Particulate Gold Occurrences in Three Carlin Carbonaceous Ores, in *Process Mineralogy VI*, ed. D. Hagni, TMS, Warrendale, PA (1986) 193-214.

D.M. Hausen and C.H. Bucknam, Study of the Robbing in the Cyanidation of Carbonaceous

Gold Ores from Carlin, Nevada, in *Applied Mineralogy*, eds W.C. Park, D.M. Hausen and R.D.Hagni, TMS, Warrendale, PA (1985) 833-856.

D.M. Hausen, C. Ekburg and F. Kula, Geochemical and XRD-computer Logging Method for Lithologic Ore Type Classification of Carlin-type Gold Ores, in *Process Mineralogy II*, ed .R.D. Hagni, TMS, Warrendale, PA (1982) 421-450.

C.L. Hayward, Cathodoluminescence of Ore and Gangue Minerals and its Application in the Minerals Industry, in *Modern approaches to ore and environmental mineralogy*, eds. L.J. Cabri and D.J.Vaughan, Mineral. Assoc. Can., Short Course, Vol 27 (1998) 269-325.

R.E. Healy, Secondary Fluorescence in Electron Microprobe Analysis Near Grain Boundaries, with Emphasis on Sphalerite and Au-Ag-Hg Alloy, submitted to *Can. Mineralogist* (2000).

R.E.Healy and W. Petruk, The Mineral Characteristics that Affect Metal Recoveries from the Cu, Zn, Pb, and Ag Ores from Manitoba, Part 1, An Investigation of the Ore Mineralogy of the Trout Lake Deposit, Mineral Sciences Laboratories, CANMET, Dept. Energy, Mines and Resources, Canada, Internal Report MSL.88-16 (IR) (1988).

R.E. Healy and W. Petruk, The Mineral Characteristics that affect Metal Recoveries from the Cu, Zn, Pb, and Ag Ores from Manitoba, Part 2, A Mineralogical Evaluation of the Behaviour of Metallic Minerals in the Trout Lake Concentrator Circuit, Mineral Sciences Laboratories, CANMET, Dept. Energy, Mines and Resources, Canada, Internal Report MSL. 89-139 (IR) (1989a) 269p.

R.E. Healy and W. Petruk, The Mineral Characteristics that affect Metal Recoveries from the Cu, Zn, Pb, and Ag Ores from Manitoba, Part 3, A Mineralogical Evaluation of the Behaviour of Metallic Minerals in the HBMS Concentrator Circuit, Mineral Sciences Laboratories, CANMET, Dept. Energy, Mines and Resources, Canada, Internal Report MSL (1989b) 150p.

R.E. Healy and W. Petruk, The Mineral Characteristics that Affect Metal Recoveries from the Cu, Zn, Pb, and Ag Ores from Manitoba, Part 4, A Mineralogical Evaluation of the Behaviour of Metallic Minerals in the Stall Lake HBMS Concentrator Circuit, Mineral Sciences Laboratories, CANMET, Dept. Energy, Mines and Resources, Canada, Internal Report MSL. (1990a) 68p.

R.E. Healy and W. Petruk, The Mineral Characteristics that Affect Metal Recoveries from the Cu, Zn, Pb, and Ag Ores from Manitoba, Part 5, A Mineralogical Evaluation of the Behaviour of Metallic Minerals in the Chisel Lake Concentrator Circuit, Mineral Sciences Laboratories, CANMET, Dept. Energy, Mines and Resources, Canada, Internal Report MSL (1990b) 90p.

R.E. Healy and W. Petruk, The Mineral Characteristics that Affect Metal Recoveries from the Cu, Zn, Pb, and Ag Ores from Manitoba, Part 6, Mineralogical Evaluation of Callinan Ores with a View to Predicting Behaviour During Beneficiation, Mineral Sciences Laboratories, CANMET, Dept. Energy, Mines and Resources, Canada, Internal Report MSL (1990c) 79p.

238

R.E. Healy and W. Petruk, Petrology of Au-Ag-Hg Alloy and 'Invisible Gold' in the Trout Lake Massive Sulfide Deposit, Flin Flon, Manitoba, *Can. Mineral.*, 28, 2 (1990d) 189-206.

R.E. Healy and W. Petruk, Graphic Galena-clausthalite Solid Solution in Low-Fe Sphalerite from the Trout Lake Massive Sulfide Ores, Flin Flon, Manitoba, *Econ. Geol.*, 87 (1992) 1906-1910.

R.S. Hedin and G.R. Watzlaf, The Effects of Anoxic Limestone Drains on Mine Water Chemistry, in *Internat. Land Reclamation Mine Drainage Conf. And Third Internat. Conf. Abatement Acidic Drainage 1*, U.S. Dept. Interior, Bureau of Mines, Special Pub. SP 06A-94 (1994) 185-194.

J.A. Herbst, Video Sampling for Mine to Mill Performance Evaluation, Model Calibration and Simulation, in Abstracts, 2000 SME Annual Meeting (2000) 62.

Hill, G.S., Application of Two-dimensional Image Analysis to Mineral Liberation Studies, Ph.D. thesis, McGill University, Dept. of Mining and Metallurgical Engineering (1990).

M.F.Hochella, B.M. Bakken and A.F. Marshall, Transmission Electron Microscopy(TEM) of Partially Oxidized Gold Ore, Carlin Mine, Nevada in *Applied Mineralogy*, eds H. Vassiliou, D.M. Hausen and J.T. Carson, TMS, Warrendale, PA (1987) 153-155.

D. Hodouin, and F. Flament, Chapter 3.1, Bilmat Computer Program for Materilas Balance Data Adjustment, in *SPOC, Simulated Processing of Ore and Coal*, ed. D. Laguitton, CANMET, Dept. Energy, Mines and Resources, Canada, CANMET Report SP85-1/3.1E (1985) 141p.

P.K. Hofmeyer and G.A. Potgeiter, The Occurrence of Psuedo-Hydrothermal Gold in Pyrite in Unisel Gold Mines Limited, *Geol. Soc. S.Africa*, Spec. Publ. 7 (1983) 271-274.

S.G. Honan and W.F. Luinstra, Gravity Gold Concentrator at Hemlo Gold Mines Inc. Golden Giant Mine, *Proceedings, 28 Annual General Meeting, Canadian Mineral Processors*, compiled by M. Mular, CIM, Mineral Processors Division, Ottawa, Canada (1996) 83-96.

X. Huang and V.P. Evangelou, Suppression of Pyrite Oxidation Rate by Phosphate Addition, in *Environmental Geochemistry of Sulfide Oxidation*, eds C.N. Alpers and D.W. Blowes, Am. Chem. Soc., Symposium Series 550 (1994) 562-573.

C.R. Hubbard and R.L. Snyder, R.I.R. - Measurement and Use in Quantitative XRD, ICDD Methods and Practices Manual, 11.2.1. (1988).

R.R. Irrinki, Key Anacon Sulfide Deposit, Gloucester County, New Brunswick, *Explor. Mining Geol.,* 1, 2 (1992) 121-129.

A.E. Isaacson and D.C. Seidel, Copper Mineral Reactions in Supercritical and Subcritical Water, in *Process Mineralogy IX*, eds. W. Petruk, R.D. Hagni, S. Pignolet-Brandom, and D.M. Hausen TMS, Warrendale, PA (1990) 343-357.

J.L. Jambor, Mineralogical Evolution of Proximal-distal Features in New Brunswick Massive-sulfide Deposits, *Can. Mineral.*, 17 (1979) 649-664.

J.L. Jambor, Mineralogy of the Caribou Massive Sulphide Deposit, Bathurst area, New Brunswick, Mineral Sciences Laboratories, CANMET, Dept. Energy, Mines and Resources, Canada, CANMET Report 81-8E (1981) 65p.

J.L. Jambor, Mineralogy of Sulfide-rich Tailings and their Oxidation Products, in *Environmental Geochemistry of Sulfide Mine-Wastes*, ed. J.L. Jambor and D.W. Blowes, Mineral. Assoc. Can., Short Course Vol. 22 (1994) 59-102.

J.L. Jambor and D. W. Blowes, Theory and Applications of Mineralogy in Environmental Studies of Sulfide-bearing Mine Wastes, in *Moderm Approaches to Ore and Environmental Mineralogy*, ed. L.J. Cabri and D.J. Vaughan, Mineral. Assoc. Can., Short Course Vol. 27 (1998) 367-401.

J.L. Jambor and J.H.G. Laflamme, The Mineral Sources of Silver and their Distribution in the Caribou Massive Sulphide Deposit, Bathurst area, New Brunswick, Mineral Sciences Laboratories, CANMET, Dept. Energy, Mines and Resources, Canada, CANMET Report 8-14 (1978) 57p.

J..L. Jambor, A.P. Sabina, R.A. Ramik and B.D. Sturman, A Fluorine-bearing Gibbsite-like Mineral from the Francon Quarry, Montreal, Quebec, *Can. Mineral.*, 28 (1990) 147-153.

J.L. Jambor and R.J. Traill, On Rozenite and Siderotil, *Can. Mineral.*, 7 (1963) 751-763.

A.P. Jeffrey and A. Joseph, Cresson Mine: Case History of a Rapidly Evolving Mining Project, *Mining Engineering*, 48, 1 (1996) 26-30.

A.E. Johnson, Mineralogical and Textural Study of the Copper-molybdenum Deposit of Brenda Mines Limited, South central British Columbia, Mineral Sciences Laboratories, Mines Branch, Dept. Energy, Mines and Resources, Canada, Mines Branch, Inf. Circular IF302 (1973).

A. Jokilaakso, R. Suominen, P. Taskinen and K. Lilius, Mineralogy and Morphology of Roasted Copper Concentrates Produced in Simulated Suspension Smelting Conditions, in *Process MineralogyIX,* eds. W. Petruk, R.D. Hagni, S. Pignolet-Brandom, and D.M. Hausen, TMS, Warrendale, PA (1990) 359-378.

Jones, M., Applied Mineralogy, A Quantitative Approach, Graham and Trotman, London, UK, (1987) 259p.

M. Kalin and R.O. van Everdingen, Ecological Engineering: Biological and Geochemical Aspects. Phase 1 Experiments, Seminar/workshops, Halifax, Nova Scotia, Canada (1987) 565-590.

M. Karakus, R.D. Hagni, S. Kang and A.E. Morris, Observation and Identification of Dross

Phases Precipitated from Lead Bullion, in *Process Mineralogy IX*, eds. W. Petruk, R.D. Hagni, S. Pignolet-Brandom, and D.M. Hausen, TMS, Warrendale, PA (1990) 415-428.

M. Karakus, R.D. Hagni and R.E. Moore, Cathodoluminescence Mineralogy of Ceramic Build-ups in Channel Induction Furnaces, in *Process Mineralogy IX,* eds. W. Petruk, R.D. Hagni, S. Pignolet-Brandom, and D.M. Hausen, TMS, Warrendale, PA (1990) 441-458.

D.A. Kepler and E.C. McCleary, Successive Alkalinity-producing Systems (SAPS) for the Treatment of Acidic Mine Drainage, in *Internat. Land Reclamation Mine Drainage Conf. and 3rd Internat. Conf. Abatement Acidic Drainage 1,* U.S. Dept. Interior, Bureau of Mines, Special Pub. Sp 06A-94 (1994) 195-204.

J.Y. Kim, S.L. Chryssoulis and K.G. Stowe, Effects of Lead Ions in Sulphide Flotation in *Proceedings, 27th Annual Meeting, Canadian Mineral Processors,* complied by L. Duval, and S. Laplante, CIM, Mineral Processors Division, Ottawa, Canada (1995) 136-154.

R.P. King, A Model for the Quantitative Estimation of Mineral Liberation by Grinding, *Int. J. Mineral. Process.* (1979) 207-220.

R.P. King and C.L. Schneider, Mineral Liberation and Continuous Milling Circuits, *Proceedings XVIII International Mineral Processing Congress* (1993) 203-211.

R.P. King and C.L. Schneider, Stereological Correction of Linear Grade Distributions for Mineral Liberation, in *Powder Technology* (1997).

S.A. Kissin and D.R. Owens, The Relatives of Stannite in the Light of New Data, *Can. Mineral.,* 27 (1989) 673-688.

R.L.P. Kleinmann, D. Crerar and R.R. Pacelli, Biogeochemistry of Acid Mine Drainage and a Method to Control Acid Formation, *Mining Eng.,* 33 (1981) 300-306.

R.L.P. Kleinmann, R.S. Hedin and H.M. Edenborn, Biological Treatment of Mine Water - An Overview, in *Proceedings 2nd Internat. Conf. Abatement Acidic Drainage 4,* MEND Secretariat, Ottawa, Canada (1991) 27-43.

R.R. Klimpel, Some Practical Approaches to Analyzing Liberation from a Binary System, in *Process Mineralogy III*, ed. W. Petruk, AIME/SME, New York, N.Y. (1984) 65-81.

H.P. Klug, Quantitative Analysis of Powder Mixtures with the Geiger-counter Spectrometer, *Anal. Chem, 25* (1953) 704.

G. Knight, T.E. Newkirk and G.R. Yourt, Full-shift Assessment of Respirable Dust Exposure, *CIM Bull.,* 67, 744 (1974) 61-72.

K. Kojonen and B. Johanson, Determination of Refractory Gold Distribution by Microanalysis, Diagnostic Leaching and Image Analysis, *Mineralogy and Petrology,* 67 (1999) 1-19.

K. Kojonen, B. Johanson, H.E. O'Brien and L. Pakkanen, Mineralogy of Gold Occurrences in the Late Archean Hattu Schist Belt, Ilomantsi, Eastern Finland, *Geol. Survey of Finland*, Special paper 17 (1993) 233-271.

K. Kojonen, P. Sorjonen-Ward, H Saarnio and M. Himmi, The Early Proterozoic Kutema Gold Deposit, Southern Finland, *Proceedings 5th Biennial SGA meeting & 10th IAGOD meeting, London, UK*, ed. C.J. Stanley et. Al., A.A. Balkema Publishers, Rotterdam, Netherlands (1999) 177-180.

J.J. Komadina and R.R. Beebe, Overcoming Climatological Limitations on Heap Leaching, in *Global Exploration of Heap Leachable Gold Deposits*, eds. D.M. Hausen, W. Petruk and R.D. Hagni, TMS, Warrendale, PA (1997) 179-187.

J. Koo and D.J. Mossman, Origin and Metamophism of the Flin Flon Stratabound Cu-Zn Sulphide Deposit, Saskatchewan and Manitoba, *Econ. Geol.*, 69 (1975) 1215-1236.

J.H.G. Laflamme and L.J. Cabri, Silver and Antimony Contents of Galena from Brunswick 12 Mine, Mineral Sciences Laboratories, CANMET, Dept. Energy, Mines and Resources, Canada, Invest. Rep. 86-138 (1986a) 13p.

J.H.G Laflamme and.L.J. Cabri, Silver and Bismuth Contents of Galena from the Brunswick No. 12 Mine, Mineral Sciences Laboratories, CANMET, Dept. Energy, Mines and Resources, Canada, Invest. Rep. 86-91 (1986b) 16p.

J.H.J. Laflamme and W. Petruk, Mineralogical Examination of Gold-bearing Samples from the David Bell Circuit, for Teck-Corona Operating Corporation, Mineral Sciences Laboratories, CANMET, Dept. Energy, Mines and Resources, Canada, Report No. MSL 91-29(CR) (1991).

D. Langmuir and D.C. Melchior, The Geochemistry of Ca, Sr, Ba and Ra Sulfates in Some Deep Brines from Palo Duro Basin, Texas, *Geochim. Cosmochim. Acta,* 49 (1985) 2423-2432.

D. Laguitton, Chapter 3.2, MATBAL Computer Program, in *SPOC, Simulated Processing of Ore and Coal,* ed. D. Laguitton, CANMET, Dept. Energy, Mine and Resources, Canada, CANMET report SP85-1/3.1E (1985) 58p.

V.I. Lakshmanan and I.G. McCool, Heap Leaching of Gold Ores in Permafrost Conditions, *Proceedings, Canadian Mineral Processors*, Compiled by G. McDonald, CIM, Mineral Processors Division, Ottawa, Canada (1987) 590-614.

A.R. Laplante, F. Vincent and W.F. Luinstra, A Laboratory Procedure to Determine the Amount of Gravity Recoverable Gold - A Case Study at Hemlo Gold Mines, in *Proceedings 28th Annual Meeting, Canadian Mineral Processors*, compiled by M. Mular, CIM, Mineral Processors Division, Ottawa, Canada (1996) 69-82.

A.C.L. Larocque, C.J. Hodgson, L.J. Cabri and J.A. Jackman, Ion-Microprobe Analysis of Pyrite, Chalcopyrite, and Pyrrhotite from the Mobrun VMS Deposit in Northwestern Quebec: Evidence for Metamorphic Remobilization of Gold, *Can. mineral.,* 33, 1 (1995) 361-372.

R. Lastra, D. Carson and D. Koren, D, Mineralogical Characterization of Leachable Elements in Ten Slags from Canadian Nonferrous Sulfide Smelters, in *Waste Characterization and Treatment,* ed. W. Petruk, SME, Littleton, CO (1998) 79-90.

R. Lastra and J.M. Greer, Preparation of Polished Sections Using Cyanoacrylate, Mineral Sciences Laboratories, CANMET, Dept. Energy, Mines and Resources, Canada, Process Mineralogy Section Report M-4408 (1992).

R. Lastra and W. Petruk, Determining Association of Unliberated Minerals in Ground Products by Image Analysis, in *Proceedings, 26th Annual Meeting, Canadian Mineral Processors*, compiled by L. Buckingham, CIM, Mineral Processors Division, Ottawa, Canada, Paper 3 (1994) 10p.

R. Lastra, W. Petruk, L.J. Cabri and J.H.G. Laflamme, Image Analysis of Samples and Microprobe Analysis of Sphalerite from the Zinc Circuit of Brunswick Mining and Smelting, Mineral Sciences Laboratories, CANMET, Dept. Energy, Mines and Resources, Canada, Division Report, MSL 95-040 (CR), (1995) 27p.

R. Lastra, W. Petruk and R. Pinard, Image Analysis of Crushed Wollastonite Drill Core Samples from B.C., Mineral Sciences Laboratories, CANMET, Dept. Energy, Mines and Resources, Canada, Division Report MSL 89-17(CR) (1989) 42p.

R. Lastra, W. Petruk and J. Wilson, Image Analysis Techniques and Applications to Mineral Processing, in *Modern Approaches to Ore and Environmental Mineralogy*, eds. L.J. Cabri and D.J. Vaughan, Mineral. Assoc. Can., Short Course Vol. 27 (1998) 327-366.

R. Lastra and R. Pinard, Image Analysis Study of Samples from University of Wales, College of Cardiff, Mineral Sciences Laboratories, CANMET, Dept. Energy, Mines and Resources, Canada, Division Report (1990) 21p.

S. LeBlanc, Massive Sulphide-associated Gold: Brunswick No.12 Mine, Bathurst, New Brunswick, Unpub. B.Sc thesis, Univ. New Brunswick, Fredericton, N.B. (1989) 51p.

P.L. LeHoux, Agglomeration Practice at Kennecott Barneys Canyon Mining Company, Barney Canyon, Ut., in *Global Exploitation of Heap Leachable Gold Deposits*, eds D.M. Hausen, W. Petruk and R.D. Hagni, TMS, Warrendale, PA (1997) 243-249.

D.P. Leta and G.H. Morrison, Ion Implantation for In-situ Quantitative Ion Microprobe Analysis, *Anal. Chem.,* 52 (1980) 277-280.

L. Lewczuk, Mineralogical Evaluation of Samples from their Grinding and Spiral Circuits of the Scully Concentrator, New Brunswick Research and Productivity Council, Report No. M/87/037, DSS contract 08200004/0001 (1988).

C.L. Lin, J.D. Miller and J.A. Herbst, An Evaluation of the Multicomponent-multisize Grinding Liberation Model Using Volumetric Grade Distribution, in *Process Mineralogy VII*, eds. A.H.

Vassiliou, D.M. Hausen and D.J.T. Carson, TMS, Warrendale, PA (1987) 589-601.

C.L. Lin and J.D. Miller, Effect of Grain Size Distribution and Grain Type on Linear and Areal Grain Distributions, in *Process Mineralogy VI,* ed. R.D. Hagni, TMS, Warrendale, PA (1986) 405-413.

G.D. Lin, A Standard Material for Liberation Analysis and Examination of the Robustness of Stereological Correction Procedures, *Ph.D. thesis, McGill University*, Dept. of Mining and Metallurgical Engineering (1996) 165p.

D. Lin, R. Lastra and J.A. Finch, Comparison of Stereological Correction Procedures for Liberation Measurements by use of a Standard Material, *Trans. I.M.M.* Section C, Sept.-Dec. (1999) C127-C137.

J.S. Livermore, The Jackling Lecture: Carlin Gold Exploration in Nevada Since the Newmont Discovery in 1961, *Mining Engineering,* 48, 6 (1996) 69-73.

J.D. Lowell and J.M. Guilbert, Lateral and Vertical Alteration-mineralization Zoning in Porphyry Ore Deposits, *Econ. Geol.,* 65 (1970) 373-408.

L. Lorenzen and J.A. Tumilty, Diagnostic Leaching as an Analytical Tool for Evaluating the Effects of Reagents on the Performance of a Gold Plant, *Mining Eng,.* 3 (1992) 503-512.

L. Lortie, W.D. Gould, M. Stichbury, D.W. Blowes and A. Thurel, Inhibitors for the Prevention of Acid Mine Drainage (AMD), in *Proceedings Sudbury 99: Mining and the Environment II,* ed. D. Goldsack, N. Belzile, P. Yearwood and G. Hall, 3 (1999) 1191-1198.

W.M. Luff, W.D Goodfellow and S.J. Juras, Evidence for a Feeder Pipe and Associated Alteration at the Brunswick No. 12 Massive Sulfide Deposit, in *Explor. Mining Geol.*, 1, 2 (1992) 167-185.

T. Lundgren and L.A. Lindahl, The Efficiency of Covering the Sulphidic Waste Rock at Bersbo, Sweden, in *Proceedings 2nd Internat. Conf. Abatement Acidic Drainage 3*, MEND Secretariat, Ottawa, Canada (1991) 239-255.

S.D. Machemer and T.R. Wildeman, Adsorption Compared with Sulfide Precipitation as Metal Removal Processes from Acid Mine Drainage in a Constructed Wetland, *Jour. Contam. Hydrol.*, 9 (1992) 115-131.

T. Malvik and B. Lund, Problems Involved with Quartzite as a Raw Material for FeSi and Si Metal Production, in *Process Mineralogy IX*, eds. W. Petruk, R.D. Hagni, S. Pignolet-Brandom, and D.M. Hausen, TMS, Warrendale, PA (1990) 499-508.

H. Mani, Determination of Characteristics of Troublesome Ultramafic Ores, DSS contract 2340-3-9266/01-SQ, Mineral Sciences Laboratories, CANMET, Dept. Energy, Mines and Resources, Canada, (1996) 32p.

M.J. Matos, Contribution for the Study of Liberation of Polymineral Ores - Techniques and Models, *Ph.D. thesis, Instituto Superior Técnico, University of Portugal*, Lisboa, Portugal (1999).

R.M.B.E. Maurice, Memoirs of Milling and Process Mineralogy: 1 - Flotation of Oxidized Ores, *Institution of Mining and Metallurgy*, Trans. Section C, 88. (1979) C245-C250.

J.A. Mavrogenes, R.D. Hagni and A.E. Morris, Reflected Light Microscopic Study of Submerged Bath Smelter Co-Ni Mattes, in *Process Mineralogy IX,* ed. W. Petruk, R.D. Hagni, S. Pignolet-Brandom, and D.M. Hausen, TMS, Warrendale, PA (1990) 399-409.

S. B. Mavrogenes and R.D. Hagni, The Mineralogy of Slags from a Secondary Copper Refinery, in *Process Mineralogy IX,* ed. W. Petruk, R.D. Hagni, S. Pignolet-Brandom, and D.M. Hausen, TMS, Warrendale, PA (1990) 379-397.

H. McCreadie, J.L. Jambor, D.W. Blowes, C. Ptacek and D. Hiller, Geochemical Behaviour of Autoclave-produced Ferric Arsenates: Jarosite in a Gold Mine Tailings Impoundment, in *Waste Characterization and Treatment*, ed. W. Petruk, AIME/SME, Littleton, CO (1998) 61-78.

S.R. McCutcheon, Base-metal Deposits of the Bathurst-Newcastle District: Characteristics and Depositional Models, *Explor. Mining Geol.*, 1, 2 (1992) 105-119.

D.C. McLean, Upgrading of Copper Concentrates by Chalcocite Derimming of Pyrite with Cyanide, in *Process Mineralogy III*, ed. W. Petruk, AIME/SME, New York, N.Y. (1984) 3-13.

G. McMahon and L.J. Cabri, The SIMS Technique in Ore Mineralogy, in *Modern Approaches to Ore and Environmental Mineralogy*, eds. L.J Cabri and D.J. Vaughan, Mineral. Assoc. Can., Short Course Vol. 27 (1998) 199-224.

W.J. McMillan, Geology and Genesis of the Highland Valley Ore Deposits and the Guichon Creek Batholith, in *Porphyry Deposits of the Canadian Cordillera*, ed. A. Sutherland Brown, CIM, Special Vol.15 (1976a) 85-104.

W.J. McMillan, J.A., in *Porphyry Deposits of the Canadian Cordillera*, ed. A. Sutherland Brown, CIM, Special Vol. 15 (1976b) 144-162.

W.J. McMillan, J.F.H. Thompson, C.J.R. Hart and S.T. Johnston, Regional Geological and Tectonic Setting of Porphyry Deposits in British Columbia and Yukon Territory, in *Porphyry Deposits of the Northwestern Cordillera of North America,* ed. T.G. Schroeter, CIM, Special Volume 46 (1995) 40-57.

S. McTavish, Goldstream Concentrator Design and Operation, *Proceedings, 17th Annual Meeting, Canadian Mineral Processors*, compiled by R. Tenbergen, CIM, Mineral Processors Division, Ottawa, Canada, paper 3 (1985) 20p.

L. Melis, K. Armstrong, A.B. Cron and R. Macphail, Metallurgy of the Mt. Milligan Porphyry Gold-Copper Deposit, in *Proceedings, 23rd annual meeting, Canadian Mineral Processors,* compiled by L. Tyreman and K. Meyer, CIM, Mineral Processors Division, Ottawa, Canada, paper 13 (1991).

Mend, Field Sampling Manual for Reactive Sulphide Tailings, Mine Environment Neutral Drainage, Report 4.1.1., CANMET, Dept. Energy, Mines and Resources, Canada (1989).

Mend, New Methods for Determination of Key Mineral Species in Acid Generation Prediction by Acid-base Accounting, Mine Environment Neutral Drainage, Report 1.16.1c., CANMET, Dept. Energy, Mines and Resources, Canada (1991).

Mend, MEND Annual Report. MEND secretariat, CANMET, Dept. Energy, Mines and Resources, Canada (1992).

J.R. Merefield, I. Stone, J. Roberts, A. Dean and J. Jones, Monitoring Airborne Dust from Quarrying and Surface Mining Operations, *Trans IMM*, 104(A) (1995) 76-78.

J..E. Micheletti and T.J. Weitz, Winter Heap Leaching at Pegasus Gold's Beal Mountain Mine, in *Global Exploration of Heap Leachable Gold Deposits*, eds. D.M. Hausen, W. Petruk and R.D. Hagni, TMS, Warrendale, PA (1997) 177.

A.R. Milnes, R.W. Fitzpatrick, P.G. Self, A.W. Fordham and S.G. Mcclure, Natural Iron Precipitates in Mine Retention Pond near Jabiru, Northern Territory, Australia, in *Biomineralization Processes of Iron and Manganese - Modern and Ancient Environments*, ed. H.C.W. Skinner and R.W. Fitzpatrick, Catena Verlag, Cremlingen-Destedt Germany (1992) 233-261.

M.M. Minnis, An Automatic Point-counting Method for Mineralogical Assessment, *Bulletin Am. Assoc. of Petroleum Geologists*, 68,6 (1984) 744-752.

D.A. Moncrieff, V.N.E. Robinson and J.B. Harris, Neutralisation of Insulating Surfaces in the Scanning Electron Microscope, *J. Physics D: Applied Physics*, 12 (1978) 2315-2325.

J.R.M. Morales, The Sotiel Mine, *Mining Magazine* (1986) 132-136.

K.A. Morin and J.A. Cherry, Trace Amounts of Siderite Near a Tailings Impoundment, Elliot Lake, Ontario, and its Implications in Controlling Contaminant Migration in a Sand Aquifer, *Chem. Geol.*, 56 (1986) 117-134.

R.D. Morrison, Applying Image Analysis to Process Control - From Blasting to Flotation, JK Centre, Australia, in *Abstracts, 2000 SME annual meeting* (2000) 41.

E. Murad, U. Schwertmann, J.M. Bigham and J. Carlson, Mineralogical Characteristics of Poorly Crystallized Precipitates formed by Oxidation of Fe^{2+} in acid Mine Sulfate Waters, in *Environmental Geochemistry of Sulfide Oxidation*, eds. C.N. Alpers and D.W. Blowes, Am.

Chem. Soc., Symposium series 550 (1994) 190-200.

D.R. Nagaraj and J.C. Brinen, SIMS Study of Adsorbed Collector Species on Mineral Surfaces. in *Proceedings XX International Mineral Process. Cong.*, eds. H. Hoberg and H.von Blottnitz, 3 (1997) 355-365.

F.W. Nentwich and R.W. Yole, Polished Thin-sections Preparation of Fine-grained Siliciclastic Rocks, *Jour. Sedimentary Petrology*, 61,4 (1991) 624-626.

H.W. Nesbitt and J.J. Muir, SIMS Depth Profiles of Weathered Plagioclase and Processes Affecting Dissolved Al and Si in Some Acidic Soil Solutions, *Nature*, 334 (1998) 336-338.

H.W. Nesbitt and J.L. Jambor, Role of Mafic Minerals in Neutralizing ARD, Demonstrated Using a Chemical Weathering Methodology, in *Moderm Approaches to Ore and Environmental Mineralogy*, eds. L.J. Cabri and D.J. Vaughan, Mineral. Assoc. Can., Short Course Vol. 27 (1998) 403-421.

H.W. Nesbitt and I.J. Muir, X-ray Photoelectron Spectroscopic Study of a Pristine Pyrite Surface Reacted with Water and Air, *Geochim. Cosmochim. Acta,* 58 (1994) 4667- 4679.

P. Neumayr, L.J. Cabri, D.I. Groves, E.J. Mikucki and J.A. Jackman, The Mineralogical Distribution of Gold and Relative Timing of Gold Mineralization in Two Archean Settings of High Metamorphic Grade in Australia, *Can. Mineral.*, 31 (1993) 711-725.

C.S. Ney, R.J. Cathro, A. Panteleyev and D.C. Rotherham, Supergene Copper Mineralization, in *Porphyry Deposits of the Canadian Cordillera*, ed. A. Sutherland Brown, CIM, Special Vol. 15 (1976) 72-78.

R.V. Nicholson, F.F. Akindunni, R.C. Sydor and R.W. Gillham, Saturated Tailings Covers above the Water Table: The Physics and Criteria for Design, in *Proceedings 2nd Internat. Conf. Abatement Acidic Drainage 1,* MEND Secretariat, Ottawa, Canada (1991) 443-460.

R.V. Nicholson, R.W. Gillham, J.A. Cherry and E.J. Reardon, Reduction of Acid Generation in Mine Tailings through the Use of Moisture-retaining Cover Layers as Oxygen Barriers, *Canadian Geotech. J.* 26 (1989) 1-8.

D.K. Nordstrom, J.W. Ball, C.E. Robertson and B.B. Hanshaw, The Effect of Sulfate on Aluminum Concentration in Natural Waters: II. Field Occurrences and Identification of Aluminum Hydroxysulfate Precipitates, *Geol. Soc. Am.*, Program Abstracts, 16,6 (1984) 611.

T. Oberthür and M. Frey, Mineralogical and Geochemical Investigation of the Ore Components from Witwatersrand-type Deposits: Their Genetic and Exploration Significance, *Proceedings ICAM 91*, ed.P.K. Hofmeyer, paper 39 (1991).

W.M. Oriel, Detailed Bedrock Geology of the Brenda Copper-molybdenum Mine, Peachland,

British Columbia, unpublished Msc thesis, Univ. of British Columbia, Vancouver, B.C. (1972).

M.J. Osatenko and M.B. Jones, Valley Copper, in *Porphyry Deposits of the Canadian Cordillera,* ed. A. Sutherland Brown, CIM, Special Vol. 15 (1976) 130-143.

D.R. Owens, Silver Distribution in Mill Products from Brunswick Mining and Smelting Corporation Limited, Bathurst, New Brunswick, Mineral Sciences Laboratories, CANMET, Dept. Energy, Mines and Resources, Canada, Invest. Rep. 80-43(IR) (1980) 23p.

A.D. Paktunc, Characterization of Mine Wastes for Prediction of Acid Mine Drainage, in *Environmental Impacts of Mining Activities,* ed. J.M. Azcue, Springer (1999a) 19-40.

A.D. Paktunc, Mineralogical Constraints on the Determination of Neutralization Potential and Prediction of Acid Mine Drainage, *Environmental Geology,* 39, 2 (1999b) 103-112.

A.D. Paktunc, J. Szymanski, R. Lastra, J.H.G. Laflamme, V. Enns and E. Soprovich, Assessment of Potential Arsenic Mobilization from the Ketza River Mine Tailings, Yukon, Canada, in *Waste Characterization and Treatment,* ed. W. Petruk, AIME/SME, Littleton, CO (1998) 49-60.

A.D. Paktunc and N.K. Davé, Acidic Drainage Characteristics and Residual Sample Mineralogy of Unsaturated and Saturated Coarse Pyritic Uranium Tailings, in *Conference Proceedings, Sudbury '99 Mining and Environment II,* eds. D. Goldsack, N. Belzile, P. Yearwood and G. Hall, Sudbury 13-17, Vol 1 (1999) 221-230.

T.F. Pederson, J.J. McNee, B. Mueller, D.H. Flather and C.A. Pelletier, Geochemistry of Submerged Tailings in Anderson Lake, Manitoba: Recent Results, in *Internat. Land Reclamation Mine Drainage Conf. And 3rd Internat. Conf. Abatement Acidic Drainage 1,* U.S. Dept. Interior, Bureau of Mines, Special Pub. Sp 06A-94 (1994) 288-296.

B. Penney, An Integrated Approach to Iron Recovery at the Iron Ore Company of Canada, in *Proceedings Annual Meeting of Canadian Mineral Processors,* compiled by M. Mular, CIM, Mineral Processors Division, Ottawa, Canada, paper 18 (1996) 213-225.

T. Peters, A Simple Device to Avoid Orientation Effects in X-ray Diffractometer Samples, *Norelco Reporter,* 17, 2 (1970) 23-24.

W. Petruk, The Clearwater Copper-zinc Deposit and its Setting, with a Special Study of Mineral Zoning around such Deposits, unpub. Ph.D. thesis, McGill University, Montreal, Quebec (1959) 122p.

W. Petruk, Aalysis of Rocks and Ores by Diffractometer, *Can. Mineral.,* 8 (1964) 68-85.

W. Petruk, Tin Sulphides from the Deposit of Brunswick Tin Mines Limited, *Can. Mineral.,*12 (1973a) 46-54.

W. Petruk, Tungsten-Bismuth-Molybdenum, Deposit of Brunswick Tin Mines Limited, its Mode of Occurrence, Mineralogy and Amenability to Mineral Beneficiation, *CIM Bull.*, 66, 732 (1973b) 113-130.

W. Petruk, Mineralogical Characteristics of an Oolitic Iron Deposit in the Peace River District, Alberta, *Can. Mineral.*, 15 (1977) 3-13.

W. Petruk, Correlation Between Grain Size in a Polished Section with Sieving Data and an Investigation of Mineral Liberation Measurements from Polished Section, *Inst. Min. and Met.*, Sec. C, 87 (1978) C272-C277.

W. Petruk, Mineralogical and Image Analysis Investigation of Iron Ore Samples from Scully Mine, Southwestern Labrador, Mineral Sciences Laboratories, CANMET, Dept. Energy, Mines and Resources, Canada, Report MRP/MSL 79-134 (IR), 1979, 62p.

W. Petruk, Mineralogy of Samples from Canada Talc, Madoc, Ontario, Mineral Sciences Laboratories, CANMET, Dept. Energy, Mines and Resources, Canada, Division Report MSL 83-123 (IR) (1983) 13p.

W. Petruk, Image Analysis System as a Monitor for Computer Control in Mineral Processing Plants, in Robotics and Automation in the Mineral Industry, CANMET, Dept. Energy, Mines and Resources, Canada, CANMET Report (1983b).

W. Petruk, Evaluation of a Middling Cleaner Tailing from the Mount Wright Concentrator by Image Analysis Studies, Mineral Sciences Laboratories, CANMET, Dept. Energy, Mines and Resources, Canada, Report MRP/MSL 84-167(IR) (1984) 16p.

W. Petruk, Image Analysis Study of the Feed, Concentrate and Tails (August, 1984) from the Carol Lake Concentrator of the Iron Ore Company of Canada, Mineral Sciences Laboratories, CANMET, Dept. Energy, Mines and Resources, Canada, Report No. 85-52(IR) (1985) 20p.

W. Petruk, Predicting and Measuring Mineral Liberations in Ores and Mill Products, and Effect of Mineral Liberations, in *Process Mineralogy VI*, ed. D.Hagni, TMS, Warrendale, PA (1986) 393-403.

W. Petruk, Ore Characteristics that Affect Breakage and Mineral Liberation During Grinding, *Applied Mineralogy VII*, eds D.J.T. Carson and A.H. Vassiliou, TMS, Warrendale, PA (1988) 181-193, also in *Proceedings 20th annual meeting Canadian Mineral Processors*, compiled by D. Loucks, CIM, Mineral Processors Division, Ottawa, Canada (1988a) 243-257.

W. Petruk, Automatic Image Analysis to Determine Mineral Behaviour During Mineral Beneficiation, *Applied Mineralogy VII*, eds. D.J.T. Carson and A.H. Vassiliou, TMS, Warrendale, PA (1988b), 347-357.

W. Petruk, The Capabilities of the Microprobe Kontron Image Analysis System, *Scanning Microscopy*, 2, 3 (1989) 1247-1256.

W. Petruk, Image Analysis of Minerals, *Short Course on Image Analysis Applied to Mineral and Earth Sciences*, ed. W. Petruk, Mineral. Assoc. of Canada, Short Course Vol. 16 (1989b) 6-18.

W. Petruk, Measurements of Mineral Liberation in Connection with Mineral Beneficiation, *Process Mineralogy IX*, ed. W. Petruk, R.D. Hagni, S. Pignolet-Brandom, and D.M. Hausen, TMS, Warrendale, PA (1990a) 31-36.

W. Petruk, Image Analysis Evaluation of Mineral Behaviour During a Laboratory Flotation Test of the Zinc-lead Volcanogenic Ore from the Faro Deposit, Yukon, *Process Mineralogy IX*, eds. W. Petruk, R.D. Hagni, S. Pignolet-Brandom, and D.M. Hausen, TMS, Warrendale, PA (1990b) 37-48.

W. Petruk, Occurrence and Recovery of Refractory Gold, *Gold 90*, ed D.M. Hausen, AIME/SME, Lidttleton, CO (1990c) 317-321.

W. Petruk, Evaluation of the Concentrator of the Iron Ore Company of Canada by Image Analysis and Materials Balancing Techniques, in *Process Mineralogy IX*, eds. W. Petruk, R.D. Hagni, D.M. Hausen and S. Pignolet-Brandom, TMS, Warrendale, PA (1990d) 111-118.

W. Petruk, Image Analysis Evaluations of Twelve Spiral Feed Samples which Produced High Weight Yields from the IOC Concentrator, Mineral Sciences Laboratories, CANMET, Dept. Energy, Mines and Resources, Canada, Job No. 50269 (1991) 36p.

W. Petruk, Preliminary Image Analysis Evaluations of Spiral Tails from the Wabush Mines' Concentrator (Comp. 1990, 1st half), Mineral Sciences Laboratories, CANMET, Dept. Energy, Mines and Resources, Canada, Division Report MSL 92-53 (IR) (1992) 15p.

W. Petruk, New Developments in the Applications of SEM/Image Analysis in Extractive Metallurgy, in *Process Mineralogy XII*, eds. W. Petruk and A.R. Rule, TMS, Warrendale, PA (1994) 173-187.

W. Petruk, The Relationship between Mineral Textures and Extractive Metallurgy, in *Process Mineralogy XIII*, ed. R.D. Hagni, TMS, Warrendale, PA (1995) 3-13.

W. Petruk, L.J. Cabri and R. Lastra, Mineralogical Evaluations of Gold Recoveries by Extractive Metallurgy Techniques, *Proceedings, 27th Annual Meeting, Canadian Mineral Processors*, complied by L. Duval and S. Laplante, CIM, Mineral Processors Division, Ottawa, Canada (1995) 412-425, also in *Process Mineralogy XIII*, ed. R.D. Hagni, AIME/TMS, Warrendale PA(1995) 191-202.

W. Petruk, G. Chung and D. Doyle, Evaluation of Spiral Performance for Processing Iron Ore from the Carol Lake Deposit of IOC, in *Proceedings, 25th Annual Meeting, Canadian Mineral Processors*, compiled by B.J.Huls and B. Brock, CIM, Mineral Processors Division, Ottawa, Canada, Paper 18, 1993, 14p.

W. Petruk, D. M. Farrell, E. E. Laufer, R. J. Tremblay, and P. G. Manning, Nontronite and

Ferruginous Opal from the Peace River Deposit in Alberta, Canada, *Can. Mineral.*, 15 (1977b) 14-21.

W. Petruk and M.R. Hughson, Image Analysis Evaluation of the Effect of Grinding Media on Selective Flotation of Two Zinc-Lead-Copper Ores, *CIM Bull.* 70, 782 (1977) 128-135, also in *Proceedings, 9th Annual Meeting, Canadian Mineral Processors*, CIM, Mineral Processors Division, Ottawa, Canada (1977) 58-76.

W. Petruk, I.B. Klymowski and G. Hayslip, Mineralogical Characteristics and Beneficiation of an Oolitic Iron Ore from the Peace River district, Alberta, *CIM Bull.* 70, 786 (1977a) 122-131.

W. Petruk, R. Lastra, J.T. Szymanski, T. Cienski, M. Beaulne and E.A. MacEachern, Mineralogical and Liberation Analyses of Graphite Ore and Mill Products for Cal Graphite, Mineral Sciences Laboratories, CANMET, Dept. Energy, Mines and Resources, Canada, Division Report MSL 92-30(CR) (1992a) 36p.

W. Petruk and R. Lastra, Evaluation of the Characteristics of Gold in the Copper Circuit of Royal Oak Mines, Mineral Sciences Laboratories, CANMET, Dept. Energy, Mines and Resources, Canada, Division Report, MSL 94-7(CR) (1994) 38p.

W. Petruk and R. Lastra, Mineralogical and Image Analysis Study of Samples from the Pb-Cu Circuit of Brunswick Mining and Smelting, Mineral Science Laboratory, CANMET, Energy, Mines and Resources, Canada, MSL-95-10, (1995) 72p.

W. Petruk and R. Lastra, Comparison of Methods of Measuring Mineral Liberations from Polished Sections, *Proceedings, 28th annual meeting, Canadian Mineral Processors*, compiled by M. Mular, CIM, Mineral Processors Division, Ottawa, Canada (1996) 509-522, also in *Proceedings EPD*, ed. G.W. Warren, TMS, Warrendale, PA (1996) 559-571.

W. Petruk and R. Lastra, Measuring Mineral Liberations and Mineral Associations from Polished Sections by Image Analysis, in *Proceedings XX Int. Min. Proc. Congress*, eds. H. Hoberg and H. Von Blottnitz, GDMB, Germany (1997) 111-119.

W. Petruk and R.D. Morris, Impurities in the Lead Concentrates of Brunswick Mining and Smelting and their Behaviour in the Lead Blast Furnace, *CIM Bull.*, 78, 878 (1984) 113-123.

W. Petruk, D.R. Owens, D.R. Morris and B. Amero, Mineralogy of Phases in the Lead Blast Furnace of Brunswick Mining and Smelting, in *Process Mineralogy III*, ed. W. Petruk, AIME/SME, New York, N.Y. (1984) 183-199.

W. Petruk, D.R. Owens and J. Szymanski, Mineralogical Analysis of Five Samples of a Garnet-kyanite Schist for Shabogamo Mining and Exploration, Mineral Sciences Laboratories, CANMET, Dept. Energy, Mines and Resources, Division Report MSL 92-37 (CR), (1992b) 21p.

W. Petruk, D.R. Owens, M.N. Sawchyn and M.M. Raicevic, Impurities in Quartzite Samples

Submitted by Shabogamo Mining and Exploration Co. Limited, Mineral Sciences Laboratories, CANMET, Dept. Energy, Mines and Rsources, Division Report MSL 92-36(CR) (1992c) 8p.

W. Petruk and R.G. Pinard, Image Analysis Study and Evaluation of Hematite in Tailings from the Mount Wright Concentrator, Quebec Cartier Mining Company, Mineral Sciences Laboratories, CANMET, Dept. Energy, Mines and Resources, Canada, Division Report No. MRP/ MSL 76-28 (IR) (1976) 19p.

W. Petruk and R.G. Pinard, Mineralogical and Image Analysis Study of Samples from Core WA20 of the Waite Amulet Tailings, Mineral Sciences Laboratories, CANMET, Dept. Energy, Mines and Resources, Canada, Report MSL 86-87(IR) (1986) 19p.

W. Petruk and R.G. Pinard, Evaluation of the Concentrator of the Iron Ore Company of Canada by Image Analysis and Materials Balancing Techniques, Mineral Sciences Laboratories, CANMET, Dept. Energy, Mines and Resources, Canada, Mineralogy Section Report 4001 (1988) 54p.

W. Petruk and R.G. Pinard, Mineralogical and Image Analysis Study of Artificially Weathered Tailings from the Midwest Uranium Ore in Saskatchewan, CANMET, Dept. of Energy Mines and Resources, Ottawa, Canada, CANMET Report 88-6E (1988) 21p.

W. Petruk, R.G. Pinard and J. Finch, Relationship between Observed Mineral Liberations in Screened Fractions and in Composite Samples, *Minerals and Metallurgical Processing*, (1986) 60-62.

W. Petruk and J.R. Schnarr, An Evaluation of the Recovery of Free and Unliberated Mineral Grains, Metals and Trace Elements in the Concentrator of Brunswick Mining and Smelting Corp. Ltd., *CIM Bull.*, 74, 833 (1981) 32-159.

W. Petruk and H.C.W. Skinner, Characterizing Particles in Airborne Dust by Image Analysis, *JOM*, 49, 4 (1997) 58-61.

W. Petruk and D.B. Sikka, The Formation of Oxidized Copper Minerals at the Malanjkhand Porphyry Copper Deposit in India, and Implications on Metallurgy, in *Process Mineralogy VII*, eds. A.H. Vassilou, D.M. Hausen and D.J.T. Carson, TMS, Warrendale, PA (1987) 403-420.

W. Petruk and J.M. Wilson, Silver and Gold in some Canadian Volcanogenic Ores, *Proceedings 8 th IAGOD Meeting* (1993) 105-117.

S.H. Pilcher and J.J. McDougall, Characteristics of Some Cordilleran Porphyry Prospects, in *Porphyry Deposits of the Canadian Cordillera*, ed. A. Sutherland Brown, C IM, Special Vol.15 (1976) 79-82.

R.G. Pinard and W. Petruk, Silver Content of Galena from Trout Lake Deposit in Flin Flon, Mineral Sciences Laboratories, CANMET, Dept. Energy, Mines and Resources, Canada,

Mineralogy Section Report M-4070 (1989).

R.G. Pinard and W. Petruk, Mineralogical Examination of Gold Samples for Teck Corona Corporation, Energy, Mines and and Resources, Mineral Sciences Laboratories, CANMET, Dept. Energy, Mines and Resources, Canada, Mineralogy Report 4301(1991) 5p.

S. Pignolet-Brandom and K.J. Reid, Mineralogical Characterization by QEM*SEM, in *Process Mineralogy VIII*, eds. D.J.T. Carson and A.H. Vassoilou, TMS, Warrendale, PA (1988) 337-346.

H. Pöllmann, The Formation and Stability of Stable Mineral Assemblages Coming from Wastes and their Probable Reuse, in *Waste Characterization and Treatment*, ed. W. Petruk, AIME/SME, Littleton, CO (1998) 149-159.

W.A. Price and J.C. Errington, ARD Policy for Mine Sites in British Columbia, U.S. Bureau Mines, Special publ. SP 06A-94, 4 (1994) 285-293.

W.A. Price, J. Errington and V. Koyanagi, Guidelines for the Prediction of Acid Rock Drainage and Metal Leaching for Mines in British Columbia: Part I. General Procedures and Information Requirements, in *Proceedings Fourth Internat. Conf. On Acid Rock Drainage 1*, MEND, CANMET, Natural Resources Canada, Ottawa (1997) 1-14.

C.J. Ptacek and D.W. Blowes, Influence of Siderite on the Geochemistry of Inactive Mine Tailings Impoundment, in *Environmental Geochemistry of Sulfide Oxidation*, ed. C.N. Alpers and D.W. Blowes, Am. Chem. Soc., Symposium Series 550 (1994) 172-189.

P. Ramdohr, Die Erzmineralien und ihre Verwachsungen. Akademie Verlag, Berlin (1975) 1277p.

E.J. Reardon and P.J. Poscente, A Study of Gas Composition in Sawmill Waste Deposits, *Reclam. Revegetation Research*, 3 (1984) 109-128.

A.J. Reed and J.L. Jambor, Highmont: Linearly Zoned Copper-molybdenum Porphyry Deposits and their Significance in the Genesis of the Highland Valley Ores, in *Porphyry deposits of the Canadian Cordillera*, ed. A. Sutherland Brown, CIM, Special Vol. 15 (1976) 163-181.

K.J. Reid and S. Pignolet-Brandom, Application of QEM*SEM for Beneficiation Studies of Minnesota Taconite, in *Process Mineralogy VIII*, eds. D.J.T. Carson and A.H. Vassoilou, Minerals, Metals and Materials Society, Warrendale, PA (1988) 369-377.

M.P. Rennick and D.M. Burton, The Murray Brook Deposit, Bathurst Camp, New Brunswick: Geologic Setting and Recent Developments, *Explor. Mining Geol.*, 1, 2 (1992) 137-142.

R.W. Rex, X-ray Mineralogy, Initial Reports of Deep Sea Drilling Project, US Government Printing Office, 4 (1970) 748-753.

G.M. Ritcey, Tailings Management, Elsevier Science Publishing Co., New York (1989) 969 p.

A.I.M. Ritchie, The Waste-rock Environment, in *Environmental Geochemistry of Sulfide Mine-Wastes*, eds. L. Jambor and D.W. Blowes, Mineral. Assoc. Can., Short Course Vol. 22 (1994a) 133-161.

A.I.M. Ritchie, Sulfide Oxidation Mechanisms: Controls and Rates of Oxygen Treatment, in *Environmental Geochemistry of Sulfide Mine-Wastes*, ed. J.L. Jambor and D.W. Blowes, Mineral. Assoc. Can., Short Course Vol. 22 (1994b) 201-244.

J.D. Robertson, G.A. Tremblay and W.W. Fraser, Subaqueous Tailing Disposal: A Sound Solution for Reactive Tailing, in *Proceedings Fourth Internat. Conf. On Acid Rock Drainage 3*, MEND, CANMET, Natural Resources Canada, Ottawa (1997) 1027-1044.

W.D. Robertson, The Physical Hydrogeology of Mill-tailings Impoundments, in *Environmental Geochemistry of Sulfide Mine-Wastes*, ed. J.L. Jambor and D.W. Blowes, Mineral. Assoc. Can., Short Course Vol. 22 (1994) 1-17.

E. Robinsky, S.L. Barbour, G.W. Wilson, D. Bordin and D.G. Fredlund, Thickened Sloped Tailings Disposal: An Evaluation of Seepage and Abatement of Acid Drainage, in *Proceedings 2nd Internat. Conf. Abatement Acidic Drainage 1*, MEND Secretariat, Ottawa, Canada (1991) 529-550.

B.W. Robinson, The "GEOSEM" (low-vacuum SEM): An Underutilized Tool for Mineralogy, in *Modern Approaches to Ore and Environmental Mineralogy*, eds. L.J.Cabri and D.J. Vaughan, Mineral. Assoc. Can., Short Course 27 (1998) 139-151.

B.W. Robinson and E.H. Nickel, A Useful New Technique for Mineralogy: The Backscattered-Electron/Low Vacuum Mode of SEM Operation, *Am. Mineral.*, 64 (1979) 1322-1328.

B.W. Robinson, N.G. Ware and D.G.W. Smith, Modern Electron Microprobe Trace-element Analysis in Mineralogy, in *Modern approaches to ore and environmental mineralogy*, eds. L.J. Cabri and D.J. Vaughan, Mineral. Assoc. Can., Short Course 27 (1998) 153-180.

A. Rosiwal, Uber Geometrische Gesteinsanalysen: Ein Einfacher weg zur Ziffermassigen Feststellung der Quantitatsverhaltnisses der Mineral-bestandteile Gemengter Gesyeine; Verh, Kaiserlich-Koeniglichen Geologischen Reichsanatait, Vienna (Translated by H.C. Ranson; On Geometric Rock Analysis. a Simple Method for the Numerical Determination of the Quantitative Ratios of the Mineral Fractions of Mixed Rocks; Royal Aircraft Establ. Farnborough, U.K., Lib. Trans. 871, 1960.) (1898) 143-75.

J.C. Rota, Mine Geology and Process Ore Control at Newmont Gold Company Mines, Carlin trend, Nevada, in *Global exploitation of heap leachable gold deposits*, eds. D.M. Hausen, W. Petruk and R.D. Hagni, TMS, Warrendale, PA (1997) 3-17.

J.T. Rybock and A.L. Anderson, Silica Micro Encapsulation: An Innovative Technology for the

Treatment of ARD, (abstract) program SME Annual Meeting, Littleton CO (2000) 54.

A.R. Sahami and D. Riehm, An Assessment of the Subaqueous Reactivity of Sulphide-rich Mine Tailings from the Petaquilla Property, Panama, *CIM Bull.* 92, 1035 (1999) 77-80.

D.F. Sangster, Precambrian Volcanogenic Massive Sulphide Deposits in Canada, A Review, Geol. Survey of Canada, paper 72 (1972).

S.C. Sarkar, S. Kabiraj, S. Bhattacharya and A.B. Pal, Nature, Origin and Evolution of the Granitoid-hosted Early Proterozoic Copper Molybdenum Mineralization at Malanjkhand, Central India, *Mineral. Deposita*, 31 (1996) 419-431.

H. Schneiderhöhn, Erzmikroskopisches Praktikum, E. Schweitzerbartshe, Verlagsbuchhandlung, Stuttgart (1952) 274p.

U. Schwertmann, Relations between Iron Oxides, Soil Color and Soil Formation, in *Soil Color*, eds. J.M. Bigham and E.J. Ciolkosz, Soil Sci. Soc. Am., Madison, Wisconsin (1993) 51-69.

G.E. Sedgwick, P.D. Dougan, J. Tremblay, R.K. Ridley and W. Jamroz, Multispectral Imager for Mine Ore Grading. *CIM Bull.* (2000) In press.

E.R Shannon, R.J. Grant M.A. Cooper and D.W. Scott, Back to Basic-- The Road to Recovery Milling Practice at Brunswick Mining, *Proceeding 25th Annual meeting, Canadian Mineral Processors*, compiled by B.J. Huls and M. Brock, CIM, Mineral Processors Division, Ottawa, Canada, paper 2 (1993) 17p.

M.L. Shutey-McCann, F.P. Sawyer, T. Logan, A.J. Schindler and I.M. Perry, Operation of Newmont's Biooxidation Demonstration Facility, in *Global exploitation of heap leachable gold deposits*, eds. D.M. Hausen, W. Petruk and R.D. Hagni, Mineral, Metals and Materials Society, TMS, Warrendale, PA (1997) 75-82.

D.B. Sikka and R.B. Bhappu, Economic Potential of Malanjkhand Proterozoic Porphyry Copper Deposit, M. P. India, *SME ,Preprint*, No. 92-160 (1992) 39p.

D. Sikka and C.E. Nehru, Review of Precambrian Porphyry Cu + Mo + Au Deposits with Special Reference to Malanjkhand Porphyry Copper Deposit, Madhya Pradesh, India, *Geol. Soc. of India*, 49,3 (1997) 239-288.

D.B. Sikka, W. Petruk, C.E. Nehru and Z. Zhang, Geochemistry of Secondary Copper Minerals from Proterozoic Porphyry Copper Deposit, Malanjkhand, India, *Ore Geology Reviews*, 6 (1991) 257-290.

I.G.L. Sinclair, Some Practical Applications of Ore Microscopy at Les Mines Selbaie, Quebec, *Process Mineralogy IX*, eds. W. Petruk, R.D. Hagni, S. Pignolet-Brandom and D. Hausen, TMS, Warrendale, PA (1990) 81-92.

H.C.W. Skinner, M. Ross and Frondel, Asbestos and other Fibrous Materials: Mineralogy, Crystal Chemistry and Health Effects, Oxford Univ. Press (1988).

H.C.W. Skinner and M. Ross, Minerals and Cancer, *Geotimes*, 14, 1 (1994) 13-15.

K.E. Smith, Cold Weather Heap Leaching Operational Methods, in *Global exploration of heap leachable gold deposits*, eds. D.M. Hausen, W. Petruk and R.D. Hagni, TMS, Warrendale, PA (1987) 189.

M.M. Smith and W. Petruk, Textural Classification Using an Image Aanalyzer, *Proceedings, Int. Symposium Production and Processing Fine Particles*, CIM/METSOC, Pergamon (1988).

M.M. Smith and W. Petruk, Applications of Fractals in Textural Analysis of Base Metal Ores, *MiCon 90: Advances in Video Technology for Microscopical Control, ASTM STP 1094*, ed. G.F. Vander Voort, Am. Soc. for Testing and Materials, Philadelphia, PA (1990).

B. Sodermark and T. Lundgren, The Bersbo Project - The First Full Scale Attempt to Control Acid Mine Drainage in Sweden, in *Proceedings Internat. Conf. Control Environment Problems from Metal Mines*, Roros, Norway (1988) 17p.

A.E. Soregaroli and D.F. Whitford, Brenda, in *Porphyry deposits of the Canadian Cordillera*, ed. A. Sutherland Brown, CIM, Special Vol. 15 (1976) 186-194.

C.J.Stanley, Optical Microscopy in Studying Ores and Process Products, in *Modern approaches to ore and environmental mineralogy*, eds. L.J. Cabri and D.J. Vaughan, Mineral. Assoc. Can., Short Course 27 (1998) 123-137.

C.J. Stanley and J.H.G. Laflamme, Preparation of Specimens for Advanced Ore-mineral and Environmental Studies, in *Modern approaches to ore and environmental mineralogy*, eds L.J. Cabri and D.J. Vaughan, Mineralogical Assoc. of Canada, Short Course Vol. 27 (1998) 111-121.

A. Stemerowitz, Personal Ccommunication, Mineral Processing Engineer, CANMET, Dept. Energy, Mines and Resources, Ottawa, Canada (1983).

R.A. Stern, High-resolution SIMS Determination of Radiogenic Tracer-isotope Ratios in Minerals. in *Modern approaches to ore and environmental mineralogy*, eds. L.J. Cabri and D.J. Vaughan, Mineral. Assoc. Can., Short Course 27 (1998) 241-268.

J.S. Stevenson, Little Billy Mines, 1944 Annual Report, Ministry of Energy, Mines and Petroleum Resources, Victoria, British Columbia (1945) A162.

M. Stichbury, G. Béchard, L. Lortie and W.D. Gould, Use of Inhibitors to Prevent Acid Mine Drainage, in *Proceedings Sudbury 95: Mining and the Environment*, eds. T.P Hynes and M.C. Blanchette, Vol II (1995) 613-622.

R.E. Stoffregen and C.N. Alpers, Svanbergite and Woodhouseite in Hydrothermal Ore Deposits: Products of Apatite Destruction During Advanced Argillic Alteration, *Can. Mineral.*, 25 (1987) 201-211.

W. E. Straszheim, K.A. Yonkin, R.T. Greer and R. Markuszewski, Mounting Materials for Automated Image Analysis of Coals Using Backscattered Electron Imaging, *Scanning Microscopy*, 2, 3 (1988) 1257-1264.

C. Sui, Z. Xu and J. Finch, Investigation of Electrochemical Aspects of Complex Sulphide Flotation - Xanthate Adsorption on Minerals as a Function of Potential - in Single and Mixed Mineral Systems. McGill Program report 2 (1996).

A. Sutherland Brown and R.J. Cathro, A Perspective of Porphyry Deposits, in *Porphyry deposits of the Canadian Cordillera*, ed. A. Sutherland Brown, CIM, Special Vol. 15 (1976) 7-16.

J. T. Szymanski and W. Petruk, Semi-quantitative and Quantitative XRD Analysis at CANMET, *Proceedings, 26th Annual Meeting, Canadian Mineral Processors*, compiled by L. Buckingham and E Roble, CIM, Mineral Processors Division, Ottawa, Canada, paper 31 (1994) 11p.

N. Tassé, M.D Germain and M. Bergeron, Composition of Interstitial Gases in Wood Chips Deposited on Reactive Mine Tailings, in *Environmental Geochemistry of Sulfide Oxidation*, ed. C.N. Alpers and D.W. Blowes, Am. Chem. Soc., Symposium Series 550 (1994) 631-634.

M.M.M.L.Tassinari, Technological Characterisation of Gold Ores. A Case Study: The Primary Ore from Salamangone, Ph.D. thesis, Mining Eng. Dept. of Polytecnic School, São Paulo University, Brazil (1996) 140p.

K.G. Thomas, G. Halverson, and K. Blower, Treatment of and Gold Recovery from Effluent at Giant Yellowknife Mines Limited, *Proceedings, Canadian Mineral Processors*, compiled by G. McDonald, CIM, Mineral Processors Division, Ottawa, Canada (1987) 479-510.

S.R. Titley and R.E. Beane, Porphyry Copper Deposits. Part I. Geological Settings, Petrology and Tectonogenesis, *Econ. Geol.*, 75th Anniversary volume (1981) 214-235.

P.C. van Aswegan, A.K. Haines and A.J. Marais, Design and Operation of a Commercial Bacterial Oxidation Plant at Fairview, *Proceedings Randol Gold Symposium, Perth, Australia*, Randol International Ltd., Lakewood, CO, USA (1988) 144-147.

E. vanHuyssteen, The Relationship between Mine Process Tailings and Pore Water Composition, in *Waste Characterization and Treatment*, ed. W. Petruk, AIME/SME, Littleton, CO (1998a) 91-108.

E. vanHuyssteen, CANMET/MMSL - INTEMIN Overview of Acid Mine Drainage in the Context of Mine Site Rehabilitation, Mineral Sciences Laboratories, CANMET, Dept. Energy, Mines and Resources, Canada (1998b) 45p.

M.W. Waldner, G.D. Smith and R.D. Willis, Lornex, in *Porphyry deposits of the Canadian Cordillera*, ed. A. Sutherland Brown, CIM, Special Vol. 15 (1976) 120-129.

J.D. Wells and T.E. Mullens, Gold-Bearing Arsenic Pyrite Determined by Microprobe Analysis, Cortez and Carlin Gold Mines, Nevada, *Econ. Geol.*, 68 (1973) 187-201.

K.D.A. Whaley, The Captain North Extension Zinc-lead-silver Deposit Bathurst District, New Brunswick. *Explor. Mining Geol.* 1, 2 (1992) 143-150.

S.A. Williams and J.D. Forrester, Characteristics of Porphyry Copper Deposits, in *Porphyry Copper Deposits of the American Cordillera*, eds. F.W. Pierce and J.G. Bolm, Arizona Geol. Soc., Digest 20 (1995) 21-34.

J.M.D. Wilson, W. Petruk and C. Coté, A Mineralogical Evaluation of the Spiral Circuit at the Mount Wright Concentrator by Image Analysis, *CIM Bull*, 83, 943 (1990) 76-83.

R.G. Wilson, F.A. Stevie and C.W. Magee, Secondary Ion Mass Spectrometry. A Practical Handbook for Depth Profiling and Bulk Impurity Analysis, John Wiley & Sons, New York (1989).

W.J. Witte and M.K. Witte, Heap Leaching in Ontario: an Example, *Proceedings, 17th annual meeting, Canadian Mineral Processors*, CIM, Mineral Processors Division, Ottawa, Canada (1985) 164-223.

J. Xu, W. Sun and S. Jia, Mineralogical and Wallrock Alteration at the Jinqingding Gold Deposit in Jiaodong Peninsula, China, *Exploration and Mining Geol.*, 3, 1 (1994) 1-8.

M.R. Woyshner and L. St-Arnaud, Hydrogeological Evaluation and Water Balance of a Thickened Tailings Deposit near Timmins, ON, Canada, in *Internat. Land Reclamation Mine Drainage Conf. And Third Internat. Conf. Abatement Acidic Drainage 2*, U.S. Dept. Interior, Bureau of Mines, Special Pub. SP 06A-94 (1994) 198-207.

E.K. Yanful, B.C. Aube, M. Woyshner and L.C. St-Arnaud, Field and Laboratory Peformance of Engineered Covers on the Waite Amulet Tailings, in *Internat. Land Reclamation Mine Drainage Conf. And Third Internat. Conf. Abatement Acidic Drainage 2*, U.S. Dept. Interior, Bureau of Mines, Special Pub. SP 06A-94 (1994) 138-147.

M. Zamalloa, T.A. Utigard and R. Lastra, Quantitative Mineralogical Characterization of Roasted Ni-Cu Concentrates, *Canadian Metallurgical Quarterly*, 34, 4 (1995) 293-301.

J. Zussman, Physical Methods in Determinative Mineralogy, Academic Press, London (1977).

SUBJECT INDEX

MINERAL INDEX